D0056928

The New York Academy of Sciences

THE NEW YORK ACADEMY OF SCIENCES STATEMENT OF PURPOSE

Science is expanding, and with it our vision of the universe. Although this new and constantly changing view may not always give us comfort, it does have the virtue of truth according to our most effective resources for acquiring knowledge. No philosophy, moral outlook, or religion can be inconsistent with the findings of science and hope to endure among educated people.

A few years ago, as part of its charge to bring new scientific information to the public, The New York Academy of Sciences launched a project to help scientists and science writers get their books published for a popular audience. Anyone who has written a book on a complex subject in the sciences has learned that it is not easy to communicate to the lay reader the excitement and enthusiasm of discovery. Science books can become technically obscure, or drift off into superficiality. Finding the balance between scientific exposition and narrative literature is a major challenge and the goal of this project.

Heinz R. Pagels
Executive Director
New York Academy of Sciences

ALSO BY PETER WARD

The Natural History of Nautilus

In Search of Nautilus

THREE CENTURIES OF SCIENTIFIC ADVENTURES
IN THE DEEP PACIFIC TO CAPTURE
A PREHISTORIC—LIVING—FOSSIL

PETER DOUGLAS WARD

A NEW YORK ACADEMY OF SCIENCES BOOK
Simon and Schuster
New York London Toronto Sydney Tokyo

Simon and Schuster
Simon & Schuster Building
Rockefeller Center
1230 Avenue of the Americas
New York, New York 10020

Published by the Simon & Schuster Trade Division

SIMON AND SCHUSTER and colophon are registered trademarks
of Simon & Schuster Inc.

Designed by Irving Perkins Associates
Manufactured in the United States of America

1 2 3 4 5 6 7 8 9 10

Library of Congress Cataloging-in-Publication Data

Ward, Peter Douglas
In search of nautilus.
"A New York Academy of Sciences book."
Includes index.
1. Nautilus. 2. Scientific expeditions—Pacific
Ocean—History. I. Title.
QL430.3.N4W36 1988 594′.52 88-11391
ISBN 0-671-61951-9

For my son,
Nicholas

Contents

Chapter 1

INTRODUCTION

For all its color, beauty, and abundance of life, it might have been the Garden of Eden. Filtered sunshine danced across the bottom of the sea, following the unusual summer storm that had blown from the north for three days. Life stirred across the sandy bottom much as it had since the miraculous several-million-year period of genesis, when the first skeletonized animals had exploded into being and populated the sea floor. The clear, blood-warm water carried its cargo of plankton, some living their short lives entirely in the upper few feet of the sea, others waiting to begin an attached, sedentary existence at the bottom. The rocks scattered across the expanses of sand were coated with encrusting animals: pavements of brachiopods, sponges, and bryozoans, endlessly filtering the plankton from the passing water. The plankton was the manna of this Eden, and virtually all creatures in this underwater garden depended on it for food.

Moving slowly, a large, crustaceanlike trilobite dredged the rippled sand, methodically ingesting fallen plankton and fecal material from the sediment. Her compound eyes scanned ahead, but as usual all she saw were the others of her kind engaged in their endless mining of the sediment. Any danger would probably come walking or crawling across the sandy sea bottom, because nothing but the harmless plankton and a few arthropodan brethren lived in the blue waters above. But danger was rare in this Eden of long-ago Arizona, and the female trilobite was well past the days of her infancy, when she too had lived in the suspended plankton; in the months following her settlement from the surface she had periodically molted and gradually grown to her foot-long mature size. Her many legs carried her, pill-bug fashion, over the sandy hummocks and hills left by the raging storm. During the storm the trilobite had remained huddled against the strong surges caused by the maddened sea surface above. Twice she had been forced to dig out of a potential grave as sand ripples marched over her. Now she fed purposefully, gathering energy and foodstuff for the eggs rapidly growing in her ovaries.

As the day wore on, the trilobite continued the endless grazing that was her entire existence. But on this day she came upon a new sight—the remains of a trilobite even larger than herself that was broken into large pieces, its entire midsection and head region shattered. A cloud of small crustaceans hovered over and on the skeleton, a testament to the recency of death. Had the trilobite possessed any cognitive power, she might have been suspicious of this scene, but of course she wasn't: She continued her endless feeding, moving past the mangled remains.

Then, on this day of firsts, she had another new experience, one that was to be her last. Suddenly she felt a constriction in her midsection, as though many ropes were suddenly tightening around her. Her universe was shattered by a cracking sound, and her middle legs went numb. More cracks followed, sending small shock waves into the endless sea as her external armature was disintegrated, leaving her soft muscles and internal organs exposed to the sea's ubiquitous, omnivorous scavengers. Her eyes looked ahead, where danger had always come from, but they saw nothing. In her last moments of life she felt something grasp her head region, then had a fleeting glimpse of long, snakelike coils emanating from a towering conical shell. Cold, Wellsian eyes examined her as death rode in on the great cracking noises. With its massive, parrotlike jaws the nautiloid cephalopod, carried into this rich garden by the storm from the north, finished breaking open the tough exoskeleton of the trilobite's head. Sinuous tentacles began passing the flesh-encrusted fragments into the mouth. The serpent had entered Eden—not along the bottom, but from a totally new direction: from the sea above. Finishing its feast, the nautiloid sucked great quantities of water into its mantle cavity, then jetted forcefully with its funnel, lifting majestically off the bottom. Now floating several feet above the sea floor, it began to swim slowly above the sand in search of its next victim, for it too was filled with developing eggs. . . .

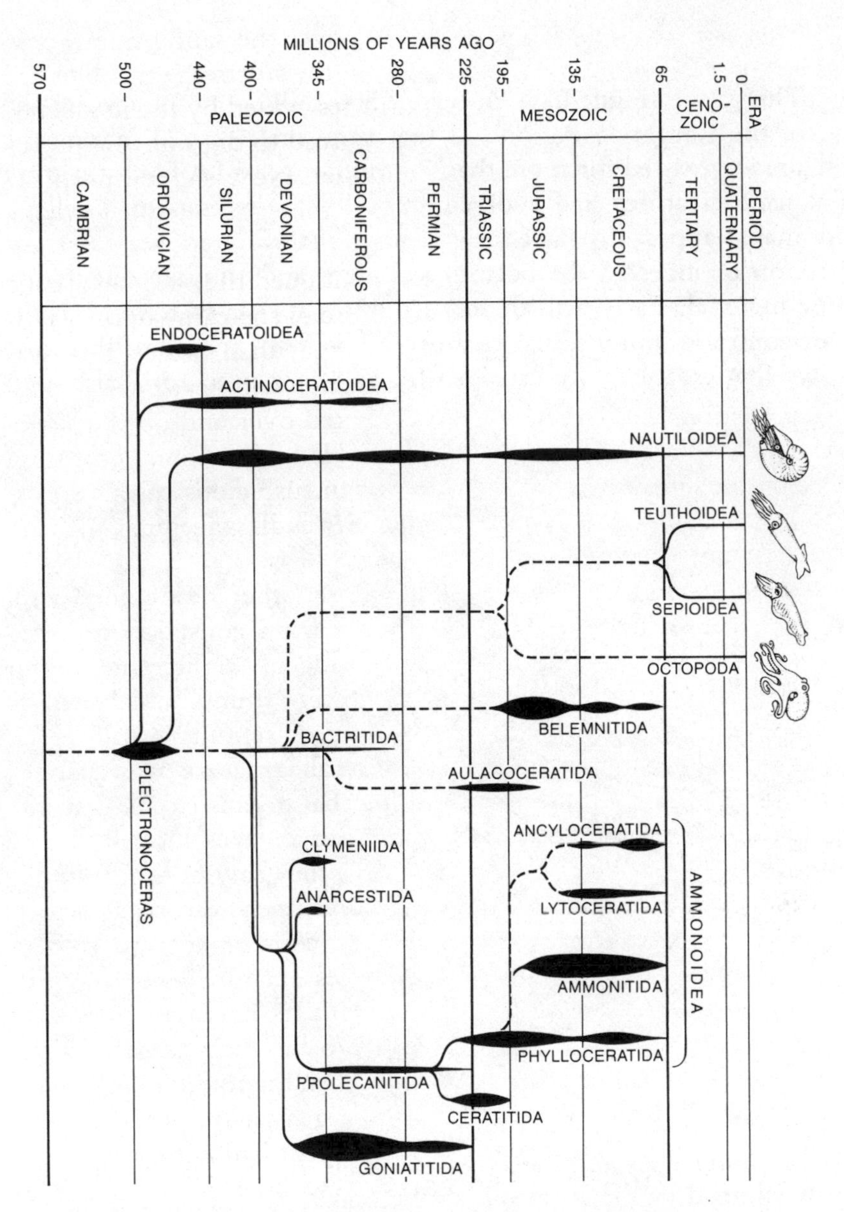

The family tree of the cephalopods. The earliest true cephalopods originated about 500 million years ago, at the end of the Cambrian Period. Three major groups of cephalopods ultimately evolved from these first ancestors: the nautiloids, of which nautilus is the last representative; the ammonoids, now extinct; and the coleoids, which include the living squid and octopus.

• • •

This scene might have occurred 500 million years ago, at the end of the Cambrian Period. At that time skeletonized marine life had already existed for more than 70 million years. A fantastic array of marine creatures had evolved, most of them bottom dwellers, and many appear to have been forms that filtered seawater for plankton or ingested the bottom sediment and strained out its organic material. Early arthropods, trilobites appear to have been the most common multicellular creatures. The fossil record of this time shows few creatures we can identify as larger predators, although

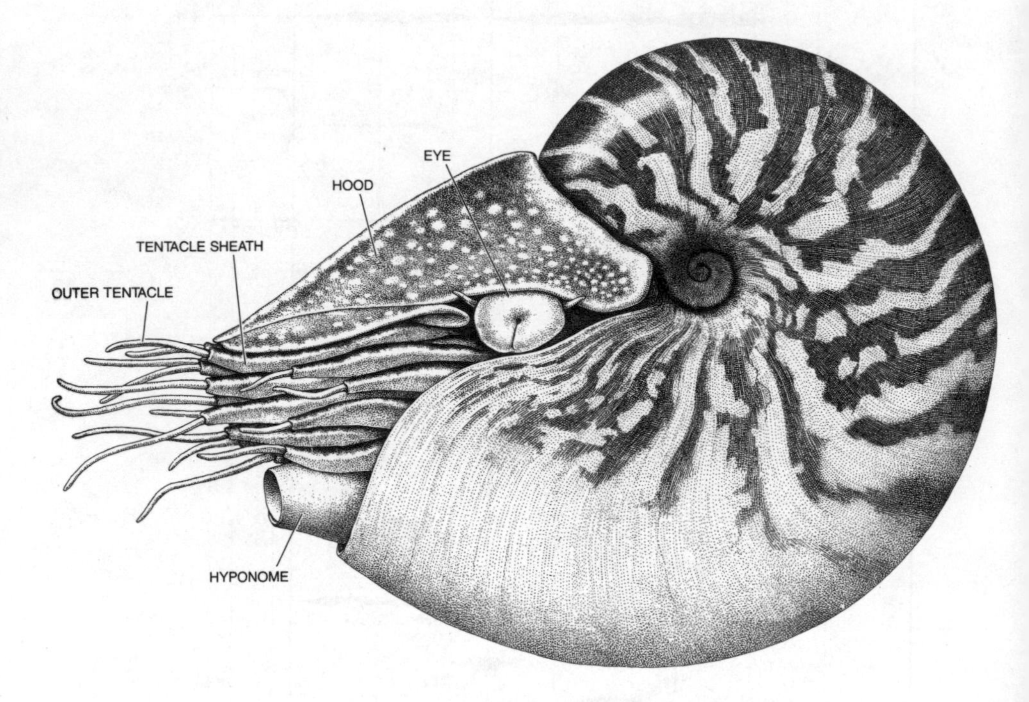

External morphology of *Nautilus macromphalus*. In life, the soft parts are mostly enclosed by the calcareous shell. Unlike those of other living cephalopods, the tentacles of nautilus are contained within sheaths. The large eyes have tiny holes instead of the true lenses other living cephalopods have. Locomotion is accomplished through the use of a *hyponome*, a funnel-like organ that extends from the bottom of the shell. Water is first drawn into the shell, passes over the gills, and then is blown out the hyponome to produce a jet propulsion that pushes the nautilus backward.

many predators without fossilizable parts may have existed. There were no fish, and no animals on the land.

Near the end of the Cambrian Period, a new evolutionary innovation occurred in what is now China. There, in sedimentary rock strata of the latest Cambrian age, small, cone-shaped shells were found among the numerous small trilobites and other fossils characteristic of this time. The shells, up to a foot long, are clearly related to the shells of other mollusks known from these and older rocks, which are ancestors of the common limpet shells and tiny clams of today. The tusklike shells found in China, however, were not only larger but also more complicated than the other shells. If one of these priceless Chinese fossils were cut open, we would find that the hind half of the shell contained many internal supports

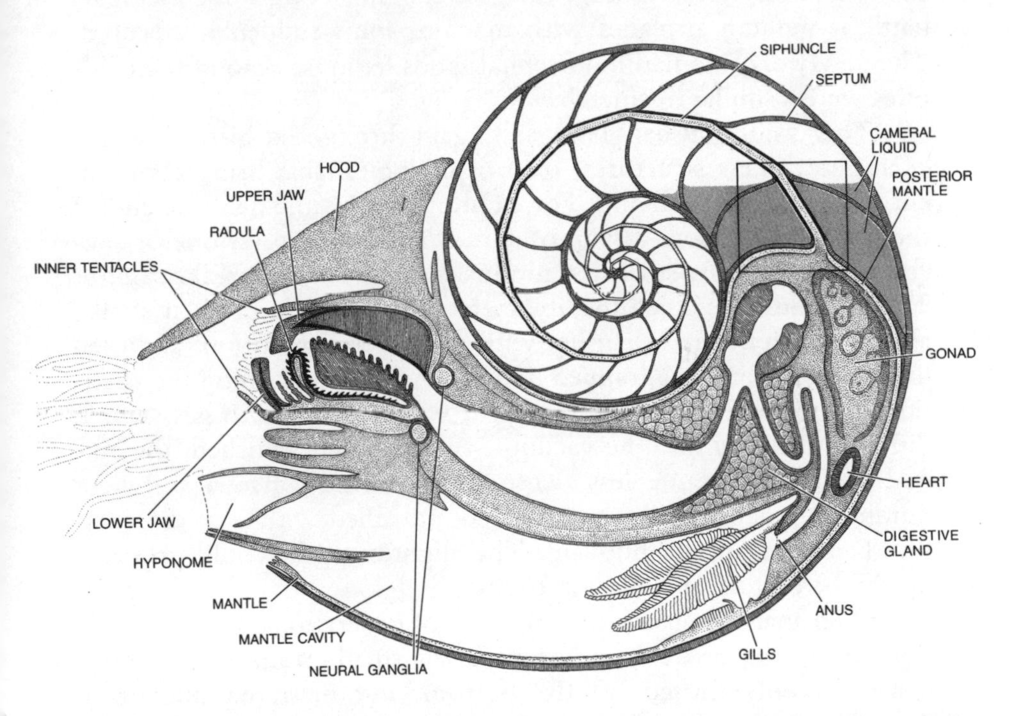

This cutaway view shows the internal organization of a nautilus. The soft parts of the body rest in the last-grown portion of the shell, and the unoccupied part of the chambered shell coils inward toward the animal's body. A central tube, called the *siphuncle,* extends back from the body and communicates with every chamber. This tube produces buoyancy change by varying the volume of liquid in each chamber.

shaped like watch crystals that divide the shell into numerous chambers. These calcareous structures, called septa, intersect the shell wall at right angles and are themselves pierced through the center by a thin tube that runs backward through every chamber. These shells are the first record of the cephalopods, a class of mollusks represented today by the octopus and squid. Today's cephalopods are all carnivorous, and in terms of speed, intelligence, and sensory ability, they represent the acme of invertebrate evolution. The Cambrian cephalopods, though less mobile and less intelligent than their descendants, must have been a devastating new predator, because they brought two new innovations to that placid world: the ability to swim and hover weightlessly above the bottom, thanks to the chambered shell, and large jaws—the first evolved—for meat eating and skeleton crushing. Imagine the impact on some medieval battle if modern airplanes with machine guns suddenly appeared. The newly evolved nautiloid cephalopods let loose among the trilobites were a similar mismatch.

This is not unusual. Time and again throughout history we see examples of new structures, or morphologies, that bring about an entirely new way of living. For example, sometime near the end of the Cambrian Period, a span of some 70 million years, one or several species of tiny, snail-like mollusks made such a breakthrough. These mollusks evolved the ability to trap liquid inside their shells, close off each flooded chamber with a calcareous partition, and then in some way force the trapped liquid out without opening the partition. As the liquid was eliminated, it was replaced with gas, which diffused passively into the vacuum. As the liquid vanished, the animal and shell became less dense overall. And when a sufficient number of chambers were evacuated and filled with gas, the creature became neutrally buoyant: The slightest push would propel it up off the bottom and allow it to float.

Over many thousands or millions of generations, natural selection refined this process, and eventually a small group of these mollusks not only floated off the bottom, but mastered the art of controlled swimming as well. The first cephalopods had evolved. Perhaps they could have become grazers or, like the great whales of today, benign floaters swimming the seas in search of plankton pastures. Instead they developed massive jaws for killing, efficient eyes for locating prey, and large size, which made them invincible. The nautiloid cephalopods became the first large, mobile predators of the sea and soon spread over the globe. They radiated explosively, evolving thousands of different species as they colonized new parts of the ocean and found new sources of prey.

These early nautiloids had long, tusklike shells that were oriented in an awkward vertical position; the gas-filled buoyancy chambers were in the apical end of the shell, while the heavier soft parts of the animal were located at the apertural end. Some species attained gigantic size, with shells 30 feet long. Others changed the awkward straight shells into coiled shapes, which gave them more maneuverability and reduced balance problems. The evolution of coiled, or planispiral, shell shapes was a major breakthrough.

Around the beginning of the Ordovician Period, about 500 million years ago, the nautiloids evolved the ability to put calcareous counterweights in the apex of their shells, thus allowing them more efficient horizontal swimming. The drawback of this solution, however, was that it made the nautiloids massive and clumsy. This was not so important during the halcyon days of the Ordovician Period, when the nautiloids were unchallenged. By the Silurian Period, however, and even more so in the Devonian Period, which started about 400 million years ago, increasing numbers of fish were appearing. For the first time the nautiloids had competition. The development of the highly maneuverable, planispiral shell shape was probably an evolutionary response to this challenge, and it allowed the nautiloids and their look-alike descendants, the ammonoids, to radiate anew.

The first appearance of nautiloid cephalopods among the late Cambrian trilobite faunas was catastrophic for the trilobites. According to fossil records, soon after the appearance of the first nautiloids a series of extinctions occurred among the trilobites; and the end of the Cambrian Period witnessed the greatest extinction among taxa of trilobites known from the stratigraphic record. Those that survived were radically different from their Cambrian ancestors. Suddenly there were trilobites with thickets of upwardly pointed spines and new, more sophisticated compound eyes that allowed stereoscopic vision in all directions simultaneously. They became less segmented and more armored. Trilobites also developed the ability to roll into a tight ball, much as a modern pill bug will do when alarmed to protect its vulnerable underside.

But in spite of these new measures, the trilobites waned in numbers over the remaining 200-million-year history of the Paleozoic Era, and gradually their place was taken by more advanced arthropods, such as the modern crustaceans, which possessed more sophisticated means of defense and predator avoidance. These new crustaceans evolved claws and the ability to rapidly flee, burrow, or roll up tightly.

As the Paleozoic Era stretched to hundreds of millions of years, the nautiloids continued to live and prosper, and continued to gorge on arthropodan flesh. Eventually the tides of time and newer innovation began to reduce the stocks of nautiloid cephalopods. Two groups of cephalopods descending from the nautiloids, the ammonoids (with shells similar to those of the nautiloids) and the coleoids (with internal shells only or no shells at all) began to take their place. By the start of the Mesozoic Era, about 225 million years ago, the nautiloid cephalopods had dwindled to a handful of planispiral genera. During the subsequent 180 million years of the Mesozoic Era, an explosion in the number of species of ammonite cephalopods, coleoid cephalopods, and fish took place, while the nautiloids neither diversified nor shrank in numbers, but remained a small group with tightly coiled planispiral shells. Sixty-five million years ago, simultaneous with the death of the dinosaurs (and the impact of comets or a large meteorite, if geochemists are to be believed), the largest stock of shelled cephalopods—the ammonoids—became extinct; the nautiloids, however, although few in number just as they had been throughout the Mesozoic Era, passed through this global catastrophe unscathed.

As more time passed, even these last nautiloid remnants dwindled in number and were no longer found worldwide. The last known fossil nautiloid cephalopod comes from rocks of the Miocene Epoch, which ended about five million years ago. Younger strata contain no known nautiloid cephalopod shells. If we were to read the fossil record literally, we would note that the 500-million-year history of the nautiloid cephalopods had finally come to an end. And today there would be no Nautilus fitness equipment, or proud submarines of fiction and fact by that name, or beautiful photos by Edward Weston showing a graceful coiled shell bisected by perfectly curved chambers. But all of these exist, as do the nautiloids. At the end of the Miocene they died out everywhere but in the tropics, where they survived in deep-water environments in front of coral reefs. Such environments have rarely left a sedimentary rock record. Nautilus, like the coelacanth—a fish that has survived extinction—has existed for the last five million years without leaving a fossil trace.

As European civilization began to explore and then conquer the faraway isles of the western Pacific Ocean, the nautiloids were rediscovered. And with discovery came study by scientists intrigued by the geometric perfection of these creatures' shell and by the evolutionary miracle of neutral buoyancy this chambered structure brought about.

Until recently, the search for nautilus required epic voyages that lasted for years. Today air travel makes the journeys far easier. I have flown across the Pacific Ocean fourteen times during the past dozen years in search of the last surviving nautiloid cephalopod. Many others have made the voyage before me, and many more, I hope, will follow. We know a great deal about this link to the past, this living fossil, but there is still more to learn. Its spiral shell is one of nature's most perfect designs and one of the reasons this animal continues to inspire scientists.

I had two major goals in writing this book. First, I wanted to show how fitful the scientific process can be, using the nautilus research as an example. Science is never a gradual accumulation of facts that always leads to the correct interpretation. Instead, it is usually full of stops and starts, and many perplexing dead ends.

As a university teacher, I have observed that one of the greatest shocks new graduate students suffer is discovering how many accumulated facts, interpretations, and generalizations their own research area depends on. And with that discovery comes the disconcerting one that they simply must trust that a certain number of their scientific ancestors were right. But scientists are human and prone to error. And sometimes these errors can go undetected for years, perhaps causing untold additional errors in subsequent work. These errors play as important a part in scientific endeavor as the discoveries, and in spite of them the young scientist must realize that he has to trust some of his forebears because no one has the time to repeat every preceding experiment or check every paper. He must, however, give that trust grudgingly.

The search for nautilus is fascinating because brave men and women have traveled to the far Pacific to study it, and sometimes they have erred. The beauty of science is that errors are usually

corrected if the problem is interesting enough to warrant continued attention. This book charts the progress and regress in the investigation of the biology of nautilus.

My second goal was to tell the story of nautilus itself. Although each chapter describes an investigator or recounts an expedition, this creature's biology will, I hope, emerge as well. We have learned much about the nautilus, but this is certainly not a complete story.

Four things about nautilus have seemed most worthy of study. First, how does the nautilus achieve buoyancy, and how does it use this buoyancy system in its everyday life? Second, where and how does nautilus live? Third, how many species are there, and what is their evolutionary history? Finally, how does nautilus reproduce and then grow, and what can its embryological history tell us about the evolution of the cephalopods? These questions inspired the voyages described in this book.

Chapter 2
THE EARLY VOYAGES

The word *nautilus* is pervasive in our language. The spiral shape of the nautilus shell is universally recognized, if not always identified. Before a bodybuilding company used the name for a new type of weight-lifting machine, the word most likely conjured up images of submarines ancient and modern. The wondrous undersea craft of Jules Verne's novel *Twenty Thousand Leagues under the Sea* was named *Nautilus,* an indication that Verne was aware of studies of the creature by comparative anatomists. These nineteenth-century scientists were trying to make sense of the nautilus shell, correctly concluding that the chambers were agents of buoyancy. For the same reason, the name *Nautilus* was also given to one of the first actual submarines, a craft built by Robert Fulton, inventor of the steamboat, at the beginning of the nineteenth century. And in more recent history—the 1950s—it was the name chosen for the U.S. Navy's first nuclear-powered submarine, part of a new generation of American underwater craft undaunted even by polar ice caps.

Nautilus shells have been part of our culture for a very long time. Beautiful chalices made from them can be found in many museums, and the word and an understanding of the animal it describes can be traced back to the ancient Greeks. But nautilus has not always been with us in the West. The last nautiluslike cephalopods to live off the shores of North America or Europe died out between five and six million years ago. The first shells known to us came back to Europe not by their own locomotion, but in the holds of ships returning from the Orient.

How and when did the first nonfossil nautilus shell make its way to Europe? The first description of it may have been made by

Aristotle, although there is some disagreement about this. The common name nautilus is today given to two entirely different animals. The pearly, or chambered, nautilus has an external shell containing air chambers. The other species, known as the paper nautilus, also produces a shell, but one that is smaller and thinner and contains no air chambers. It is produced by a small octopus properly called *Argonauta* that is well known in the Mediterranean Sea. Aristotle described argonauts in his *Historia animalium*.

One passage in Aristotle's work in which he describes the animals we now call cephalopods (one of the major classes of mollusks) refers to a creature unlike any of the Mediterranean cephalopods he would have been familiar with. Here Aristotle compares a strange shelled form with the other, better-known cephalopod species: "But the other genus is in a shell, like a snail; it never quits its shell, but exists after the manner of a snail, and sometimes outwardly extends its arms." It is the last part of this sentence that is most enigmatic. Among the cephalopods, nautilus alone has tentacles that retract or extend from a series of covering sheaths. The only other possible candidate for Aristotle's description, the argonaut, has different tentacles and can completely leave its shell, whereas a nautilus is permanently attached within. Richard Owen, the great nineteenth-century comparative anatomist who translated this critical sentence, was sure that Aristotle was actually making the first description of the soft parts of nautilus.

But this question remains: How could Aristotle have seen a living or preserved specimen of nautilus? This last relic of the ancient nautiloids is found mainly in the western Pacific Ocean, an area never explored by the ancient Greeks or their proxies. Even the far-roving Alexander the Great never visited areas of the southwestern Pacific basin inhabited by nautilus. For this principal reason, historians have doubted that Aristotle was acquainted with the chambered nautilus.

Not everyone is convinced by such arguments. The zoologist Anna Bidder, an expert on the biology of nautilus, has studied the relevant passages and is convinced that Aristotle's description could refer only to nautilus. Moreover, Bidder believes that Aristotle must either have seen a living specimen or at least talked with someone who had. She suggests that Aristotle could have accomplished this only one way: His nautilus came from the eastern coast of Africa.

The ancient Greeks were certainly familiar with many areas of Africa, and there is good reason to believe that nautiluses live there today, because their shells are common along the coast of Kenya. Some of these shells are clean and free of epizoan invertebrate life, which suggests they have not spent long periods of time drifting. Nautiluses are also known to live in several nearby localities in the Indian Ocean. This evidence makes it very likely that populations of nautilus exist along much of the coast of eastern Africa. Another possibility is that a stray specimen with the living animal still inside drifted to the African coast from a population living farther to the east, such as the western coast of Australia; this theory is plausible in that a living nautilus recently found off Japan had drifted thousands of miles from its home in the Philippines. In either case, a Greek explorer could have seen a nautilus in Africa and described or brought the animal to Aristotle.

Not much else was written about nautilus until the Renaissance. According to Dr. Richard Davis of the Cincinnati Museum, who has recently written an excellent review of the early history of nautilus investigation, the first modern use of the name nautilus, in the mid-1500s, was by a Frenchman named Pierre Belon. In Belon's time nautilus shells became relatively common in Europe as increasing numbers of explorers and merchants opened up trade routes to the east. According to Anna Bidder, the nautilus shell was one of the three great natural treasures of that time, along with the ostrich egg and the coconut. These curios were often turned into ornate chalices; many exquisite pieces constructed around nautilus shells are known from this period.

Nautilus changed from a curio to an object of scientific inquiry in the mid-1600s, largely because of the efforts of two men: the great English scientist and philosopher Robert Hooke, and a Dutch naturalist named Georgius Everhardus Rumphius.

Hooke, reputed to have been one of the most intelligent men who ever lived, was hired by the Royal Society of London to make a discovery a day. Had he not been immediately succeeded in his post by the great Isaac Newton, he would probably be far better known today. Hooke and the younger Newton detested each other, and after Hooke's death, Newton did everything possible to downplay his predecessor's accomplishments. A naturalist, architect, scientist, and Renaissance man, Hooke used his probing intellect to examine all manner of things, from the microscopic to the cosmic.

And nautilus certainly did not escape his gaze; on three different occasions he spoke about it to the Royal Society of London.

Hooke was a remarkably perceptive man. According to Dr. Eric Denton, whose own pioneering work on nautilus will be addressed later in this book, Hooke was the first to realize that the nautilus shell and animal were connected to each other by a tubular structure that is found in each chamber. This structure, called the siphuncle, consists of an outer calcareous and chitin tube and an inner section of vascularized flesh composed of living tissue. It is the key to buoyancy regulation by nautilus. Hooke said of this tube: "This admirable structure seems to me not a mere Lusus Naturae, or a Form by Chance, to express a Variety, but an Emanation of that infinite Wisdom, that appears in the Shapes and Structure of all other created Beings, which is to endow them with sufficient Abilities to perform those Actions, which are made necessary to their Well being."

Hooke observed spacious chambers in the nautilus shell and clearly realized that by regulating their contents, the creature could reduce the density of the entire organism. He also noted how important the ability to manage this buoyancy mechanism would be to the nautilus. Hooke saw that the most efficient method of changing buoyancy would be to add water to or remove it from the chambered parts of the shell. What he had to figure out was a reasonable explanation for how this could be accomplished. Hooke was a contemporary of Robert Boyle, whose name is lent to one of the great laws of nature that describes properties of gas volumes under varying pressures. Getting water into a nautilus shell at depth would be no problem, but removing that water from the chambers was another story. Here is how Hooke supposed that process might work:

> It was easy to conceive, how it could fill his shell with Water, and so sink himself to the Bottom; but then how (when there is such a Distance, from the Air) he could evacuate the Water, and fill the Cavities with Air, that was difficult to comprehend, especially being under so great a Pressure of Water: But if nature had furnished him with a Faculty of producing artificial Air, then the Riddle would be quickly unfolded. I found, therefore, that by Art it was feasible to produce such an artificial Air, and that it was endured with a very great Power of Expansion, so that it would not only make itself Room to expand, notwithstanding the incumbent Pressure of the Air on all Sides; but, if sealed up in strong Glasses, it would break out the Sides thereof,

> which might have as much Power of Expansion as might counterpoise, nay, outpower both the Pressure of the Air, and also the Water too, though 100 times greater than the Air.

Hooke was describing the production of gas under pressure by a biological tissue. Such a procedure seems to be the most logical way of explaining the removal of water from the chamber of a nautilus; exactly such a system is used in modern submarines to blow ballast and allow ascent. We know that biological systems are capable of such a feat. The very great success of the modern fishes is in no small way related to their ability to manufacture high-pressure gases in a small organ called the swim bladder. But is such a system used by cephalopods as well as fish? The swim bladder of a fish is a specialized and clearly complicated organ; Hooke thought that nautilus must possess something similar. His suggestions about the means by which nautilus achieves buoyancy created an avenue of inquiry that continues to the present day, and started a general search for the creature's gas-producing organ.

Hooke made another great intuitive jump. Being a well-rounded naturalist, he was aware of the great diversity of fossil forms resembling nautilus. These fossil nautiloids and ammonoids, mainly Mesozoic in age, can still be mined out of sedimentary strata in many areas of England. Hooke decided that the fossil forms probably would have employed the same buoyancy mechanism used by the modern nautilus. But his speculations were based on studies of the shell only, because the soft parts associated with these shells were virtually unknown. The first published description of the soft parts appeared in 1705 in a work by Georgius Rumphius, a Dutchman living in the Dutch East Indies—today's Indonesia. Rumphius's illustrated descriptions of nautilus's soft parts were not especially accurate. From their general plan, however, it was clear to other early anatomists that nautilus was very different from the other cephalopods. For example, Rumphius's illustrations showed more than the ten tentacles that were common to most cephalopods. For an accurate picture to emerge, it was vital that well-preserved soft parts become available to anatomists in Europe—or that these anatomists travel to where the nautilus lived.

Comparative anatomy became a viable scientific discipline during the late eighteenth century. By describing in minute detail the anatomy of various organisms, and then comparing them one to another, early scientists found a method of deducing taxonomic relationships. With the incorporation of Darwin's theory of evolution in the middle of the nineteenth century, the last piece in the puzzle of similarities and differences among organisms was in place.

Richard Owen was a pioneer in the study of comparative anatomy. An Englishman who mastered the anatomical details of most vertebrates and invertebrates, he is known for his classic studies of many fossil vertebrates from his native country. His first major study, however, was of a contemporary nautilus brought from the distant Pacific.

Owen had the great fortune of acquiring the first preserved soft parts of a nautilus ever to come to Europe. He received this invaluable specimen from a George Bennett, who had recently returned from a voyage to the far Pacific. In 1829 Bennett, a member of the Royal College of Surgeons, was aboard one of His Majesty's ships visiting the New Hebrides* (now Vanuatu), an island group to the east of Australia. These islands were a particularly dangerous place to explore, as cannibalism was still a common practice. In his diary, Bennett documents his capture of this first nautilus:

> Island of Erromanga, New Hebrides, August 24, 1829. Monday; fine weather during the day. Thermometer at noon, 79. In the evening a Pearly Nautilus (*Nautilus pompilius* Linnaeus) was seen in Marekini Bay, on the southwest side of the island, floating on the surface of the bay, not far distant from the ship, and resembling, as the sailors expressed it, a dead tortoise-shell cat in the water. It was captured, but not before the upper part of the shell had been broken by the boat hook in the eagerness to take it, as the animal was sinking when caught. On its being brought on board, my attention was directed to possessing the inhabitant, which I succeeded in procuring.

The species found in the New Hebrides today is assigned to *Nautilus pompilius*. From Bennett's description we get the impression that the nautilus he obtained was alive, because he says that it began to sink when the men approached it; probably it was swimming downward. If the specimen had been dead, it could not have

*In order to maintain a sense of continuity with the earlier source material quoted throughout this book, the author has retained the European names for Vanuatu (*New Hebrides*) and Belau (*Palau*).

sunk until *after* the shell was broken by the overzealous sailors. The New Hebrides is, along with nearby New Caledonia and the Loyalty Islands, one of the few areas where nautilus can be encountered in shallow water at night. For this reason Bennett's was possibly a healthy specimen rather than an injured or moribund form.

Bennett made one more highly interesting observation. He spoke of finding *liquid* within the chambers of the shell; Hooke also had hypothesized the presence of chamber liquid, and we now know that nautilus maintains liquid of varying volumes there. Bennett noted that his specimen had been broken open with a boat hook while being recovered from the sea, and this accident would certainly have allowed seawater into some of the chambers. But Bennett's passage gives an impression of naturally contained liquid: "On laying open that portion of the shell which contains the chambers, it was found to contain water, which of course immediately escaped." Surely Bennett would have differentiated the chambers he opened himself from those already broken during recovery of the animal. Does this passage mark the first observation of chamber liquid, which plays such an important role in the buoyancy and growth of nautilus?

Richard Owen received Bennett's nautilus, which had been preserved in "spirits," from the Royal College of Surgeons in 1831. He immediately began studying it and published a monograph describing its anatomy in 1834. Owen's descriptions of the nautilus's various anatomical systems was highly detailed. It is remarkable that he was able to do so much with this single, imperfectly preserved specimen.

Owen made many original observations of both the shell and the soft parts. He was the first to note a thin sheet of membrane lining the walls of the chambers, a structure he called the pellicle, the name still used today. The pellicle is involved in the removal of liquid from the chambers of a nautilus, and therefore plays a key role in the formation and maintenance of neutral buoyancy. At the time of Owen's work, it was believed that a similar membrane had existed in fossil specimens, and that, as Hooke had supposed, it was the organ responsible in both recent and fossil forms for secreting gas. According to this idea, the tubular siphuncle contained blood veins that kept the pellicle alive in the chambers. Owen appears to have been the first to doubt that this particular system could function as supposed. In discussing Hooke's theory, Owen wrote:

> The above appearances [of pellicle] in the fossil shells have been deemed confirmatory of the hypothesis of Dr. Hooke, who supposed *Nautilus pompilius* to have the power of generating air into, and expelling it from, the deserted chambers; and that it regulated, in the same manner as Fish by means of their air-bladders, its ascent and descent in the water.

But Owen took into account the sizes of both membrane and arterial systems and concluded that the siphuncle was too small to perform the necessary function:

> . . . The size of the artery seems barely adequate to support the vitality of the membrane, much less to effect a secretion, for which in a fish an ample gland appears to be indispensable; and with respect to the outlet [the place where the end of the siphuncular tube communicates into the main part of the shell], the oblique and contracted nature of the passage is ill-calculated to allow an escape of gas sufficiently rapid to answer as a self-preserving action, or a means of defense against sudden assaults.

Owen's observations were prescient and correct. The diameter of the siphuncle is far too small to allow the movement of gas in sufficient amounts to change buoyancy as earlier hypothesized. Unfortunately, many subsequent observers ignored this judgment and blithely followed Hooke in searching for the origin of artificial air.

The nautilus's mode of life, especially how it used its chambers, was very much on Owen's mind. Owen wanted to test his ideas about buoyancy and the chambered shell portion on a living nautilus. About this he wrote:

> Much, indeed, remains to be done before the theory of the chambers and siphuncle can rest on the sound basis of experimentation and observation. Mr. Bennett's observation—that the contents of the deserted chambers in the living animal are liquid—is an important addition to its history; though it may still be doubted whether their contents are the same under all circumstances, even during the lifetime of the animal; and the nature of the fluid, its proportional quantity, and the precise disposition and contents of the membranous tube, still remain to be determined.

Owen also described a crucial experiment that could be carried out to determine the nature of gas pressure within the chambers:

> It would be advisable, in the event of another fortunate capture of the nautilus, to lay open the chambers under water, when the presence of the gas in any of them would be ascertained, and it might be received: the contents of the central tube, if gaseous, would at once be detected.

With these comments Owen showed the way for the next studies of nautilus. However, the next generation of explorers, such as Bashford Dean and Arthur Willey, apparently ignored his call to more closely examine the nature of the chambers and their contents. This oversight impeded the correct interpretation of chamber formation and buoyancy maintenance for almost a century.

At the end of the nineteenth century, nautilus was still an enigmatic creature. Although comparative anatomists had described the soft parts in some detail, many other aspects about the animal—its mode of life, habitat, reproductive habits—remained shrouded in mystery. The animal's buoyancy was especially puzzling. Ray Lankester, one of the greatest zoologists of the late nineteenth century, summed up the state of knowledge about nautilus's buoyancy and chamber formation for the ninth edition of the *Encyclopaedia Britannica:*

> There appears to be no doubt that the deserted chambers of the nautilus shell contain in the healthy living animal a gas which serves to lessen the specific gravity of the whole organism. The gas is said to be of the same composition as the atmosphere, with a larger proportion of nitrogen. With regard to its origin we have only conjectures. Each septum shutting off an air-containing chamber is formed during a period of quiescence, probably after the reproductive act, when the visceral mass of the nautilus may have slightly shrunk, and gas is secreted from the dorsal integument so as to fill up the space previously occupied by the animal. A certain stage is reached in the growth of the animal when no new chambers are formed. The whole process of the loosening of the animal in its chamber and of its slipping forward when a new septum is formed as well as the mode in which the air chambers may be used as a hydrostatic apparatus, and the relation to this use, if any, of the siphuncular pedicle, is involved in obscurity, and is the subject of much ingenious speculation.

Although this passage may have been speculative, our present knowledge shows that it was not very ingenious. The rather Victorian assumption that the tired animal would produce new chambers after the reproductive act is untrue; the nautilus does not begin reproduction until all of the chambers are finished. The idea that gas is produced in a space behind the body and then sealed off with a new septum was also greatly in error. But Lankester was powerful enough among the scientists of his day that his comments shaped the consensus about nautilus for several generations.

Further understanding of the nautilus clearly required observation of living specimens. It was time for determined scientists to go out into the Pacific.

Chapter 3

ARTHUR WILLEY IN THE PACIFIC WILDERNESS, 1894–1897

The detailed anatomical studies of Owen and other comparative anatomists of his day revealed much about nautilus. But sadly, specimens available for study were scarce, and those few that did exist were usually poorly preserved. Many questions thus remained about nautilus. Perhaps the greatest mystery of all concerned its embryonic development. Embryology had not yet been carefully studied as a branch of biology, but as the nineteenth century came to a close, zoologists were increasingly aware of its importance. By tracing each step in the development of an embryo from fertilization to birth, they could discover major pathways of evolution in general. Because of the great antiquity of the nautiloids and their central position as originators of all other cephalopods, understanding their embryological history was clearly a top priority. However, it was also clear that an investigator would have to visit the nautilus's habitat to obtain such an embryological series.

The first to take this up was Arthur Willey, a young Englishman studying at Cambridge University. Of all the voyages made in search of nautilus, Willey's was the most epic. He left England in 1895 at the age of twenty-eight and did not return from the Far East for three years. His mission was simple: Obtain specimens documenting the development of nautilus, as well as any other information he considered interesting. Although Willey failed to bring back the embryological series so coveted by the Victorian anatomists, his voyage and its results (documented in a thick monograph published in 1902, one of seven compiled from information or specimens he collected) provided tantalizing glimpses about the ecology, physiology, and evolution of this living fossil. Willey's work is still the most important source of information about the anatomy of nautilus and continues to be an essential stepping-stone for all modern studies.

WILLEY AT CAMBRIDGE

Little is known about Willey's origins and youth. However, we know a great deal about his voyage to the South Pacific from his monograph *Zoological Results, based on material from New Britain, New Guinea, Loyalty Islands, and Elsewhere*. In a heroic style now out of fashion, Willey recounted not only his scientific pursuits, but tales of the peoples, islands, and experiences he encountered during his three-year journey. This work leaves us with a vivid impression of an otherwise quite elusive man. He attained the position of Fellow of the Royal Society in 1902 and accordingly was given the impressive obituary that all FRS's receive in *The Times* of London, but his life was never the subject of a single biography; he is, however, now often mentioned in the numerous compendiums of the lives of leading scientists.

Born in 1867 in Scarsdale, England, Willey was educated in Bath before attending University College, London. There he was influenced by the great English zoologist Ray Lankester, a specialist in invertebrate zoology and longtime director of the British Museum. After completing his bachelor's work, Willey became a tutor at Columbia University in New York, where he wrote a book on the origin of the vertebrates. In 1894, on Lankester's recommendation, Willey was named Balfour Fellow at Cambridge. Also at Lankester's suggestion, the young man abandoned his vertebrate studies to begin his quest for the Holy Grail of invertebrate embryology: the developmental sequence of nautilus.

Lankester appears always to have been close in Willey's thoughts during his three-year voyage, and his influence on matters pertaining to nautilus may have been critical. Lankester had written the influential summary of what was then known about nautilus for the *Encyclopaedia Britannica*. This treatise described the way new chambers were formed and how neutral buoyancy was achieved. As we shall see, Willey followed his master's lead in these matters without question, even in the face of overwhelming evidence contradicting Lankester's ideas. In this one respect, Willey the exquisite observer became Willey the blind. This crucial mistake was not be be corrected for nearly 70 years.

WILLEY IN NEW BRITAIN

Thanks to his marvelous descriptions of the people and places he visited, we have a great deal of information about Willey's voyage. The outward journey took him from London through the Mediterranean and Suez to Singapore. On the way he stopped in Naples, where he conferred with the great zoologist Anton Dohrn. Dohrn, a leading student of embryology, must have been as excited as Lankester was about the prospect of finally getting the embryological series of nautilus.

Willey's first destination was New Britain, a large island in the Bismarck Archipelago, located to the northeast of Papua New Guinea. While still in England preparing for his trip, Willey had learned from missionaries just arrived from New Britain that a local native fishery for nautilus already existed there. Based on this information, he decided to begin his studies around the Gazelle Peninsula of New Britain. After changing ships in Singapore, Willey sailed into the clear waters of Blanche Bay, New Britain, sometime in mid-December 1895. His first view of his new home was dominated by the ominous profiles of The Mother and Daughters, volcanoes that had been active intermittently throughout the nineteenth century and were active again at the time of Willey's visit.

Willey's top priority after his arrival was to find a boat, and his second was to find lodgings. He got his first boat through the good offices of an English missionary named Richard Helfer. Two days after delivering the boat to Willey, the priest took his own boat from the town of Ralum to the nearby island of Matupi, sheltered in the large confines of Blanche Bay. Helfer was engaged in routine missionary work, searching for Christmas ornaments, when a sudden squall came up during his seven-mile trip. His boat foundered in the rising waves and he was lost. Willey spares us the gory details, but it is more than likely that Helfer was eaten by sharks. Even in high wind the warm, salty seawater of this region makes drowning unlikely, especially if a floating (if swamped) boat is available. But the sharks in that part of the world are unbelievably aggressive.

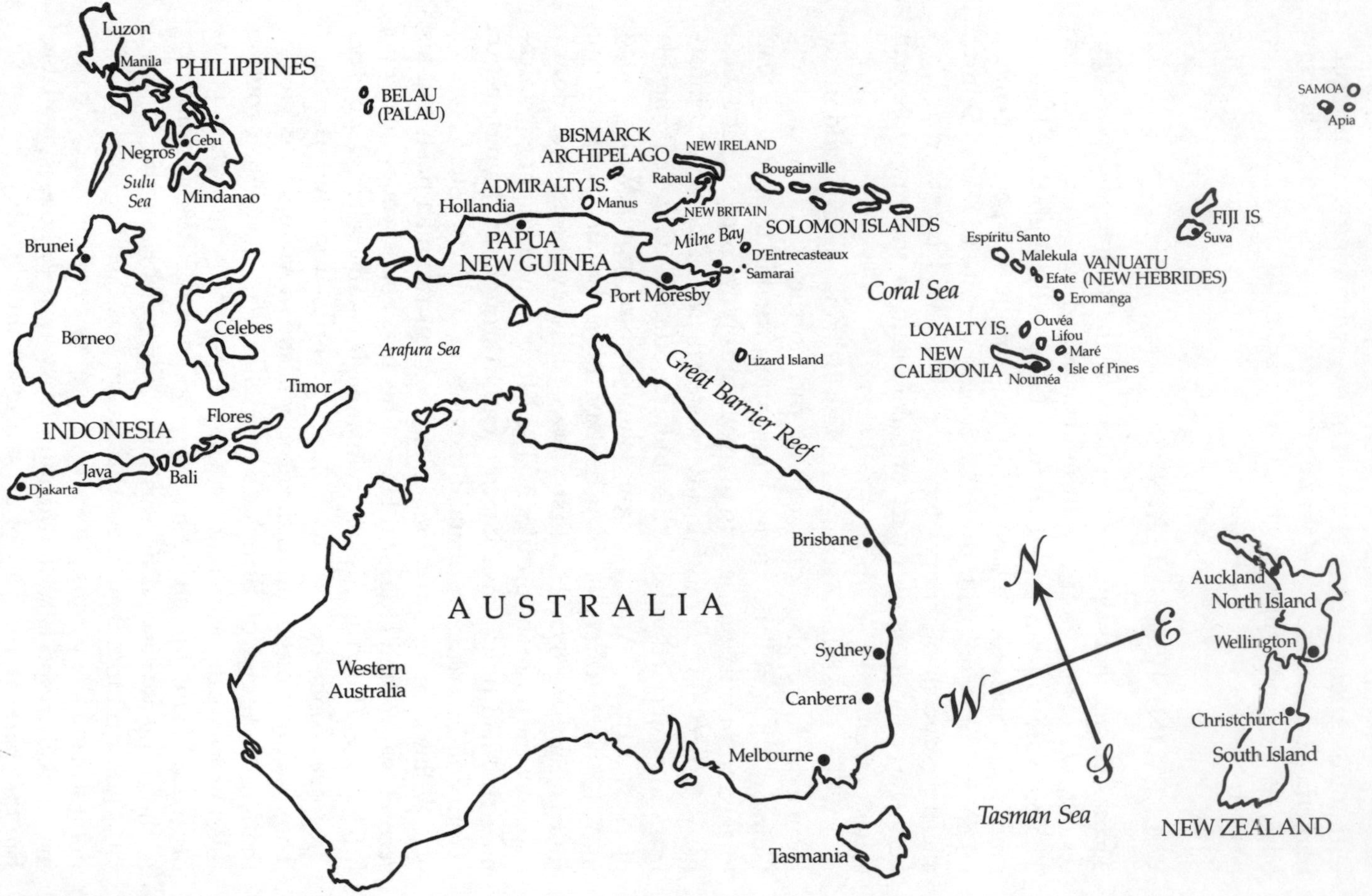

Luzon
Manila
PHILIPPINES
BELAU
(PALAU)
SAMOA
Apia
Cebu
Negros
Sulu
Sea
Mindanao
BISMARCK
ARCHIPELAGO
NEW IRELAND
ADMIRALTY IS.
Rabaul
Bougainville
Manus
Hollandia
NEW BRITAIN
SOLOMON ISLANDS
FIJI IS
Suva
Brunei
PAPUA
NEW GUINEA
Milne Bay
D'Entrecasteaux
Samarai
Espíritu Santo
Malekula
Efate
VANUATU
(NEW HEBRIDES)
Eromanga
Port Moresby
Coral Sea
Borneo
Celebes
Arafura Sea
LOYALTY IS.
Ouvéa
Lifou
Maré
Isle of Pines
NEW
CALEDONIA
Nouméa
Lizard Island
Great Barrier Reef
Timor
Flores
INDONESIA
Java
Bali
Djakarta
Brisbane
AUSTRALIA
N
E
W
S
Auckland
North Island
Wellington
Sydney
Western
Australia
Canberra
Christchurch
South Island
Melbourne
Tasman Sea
NEW ZEALAND
Tasmania

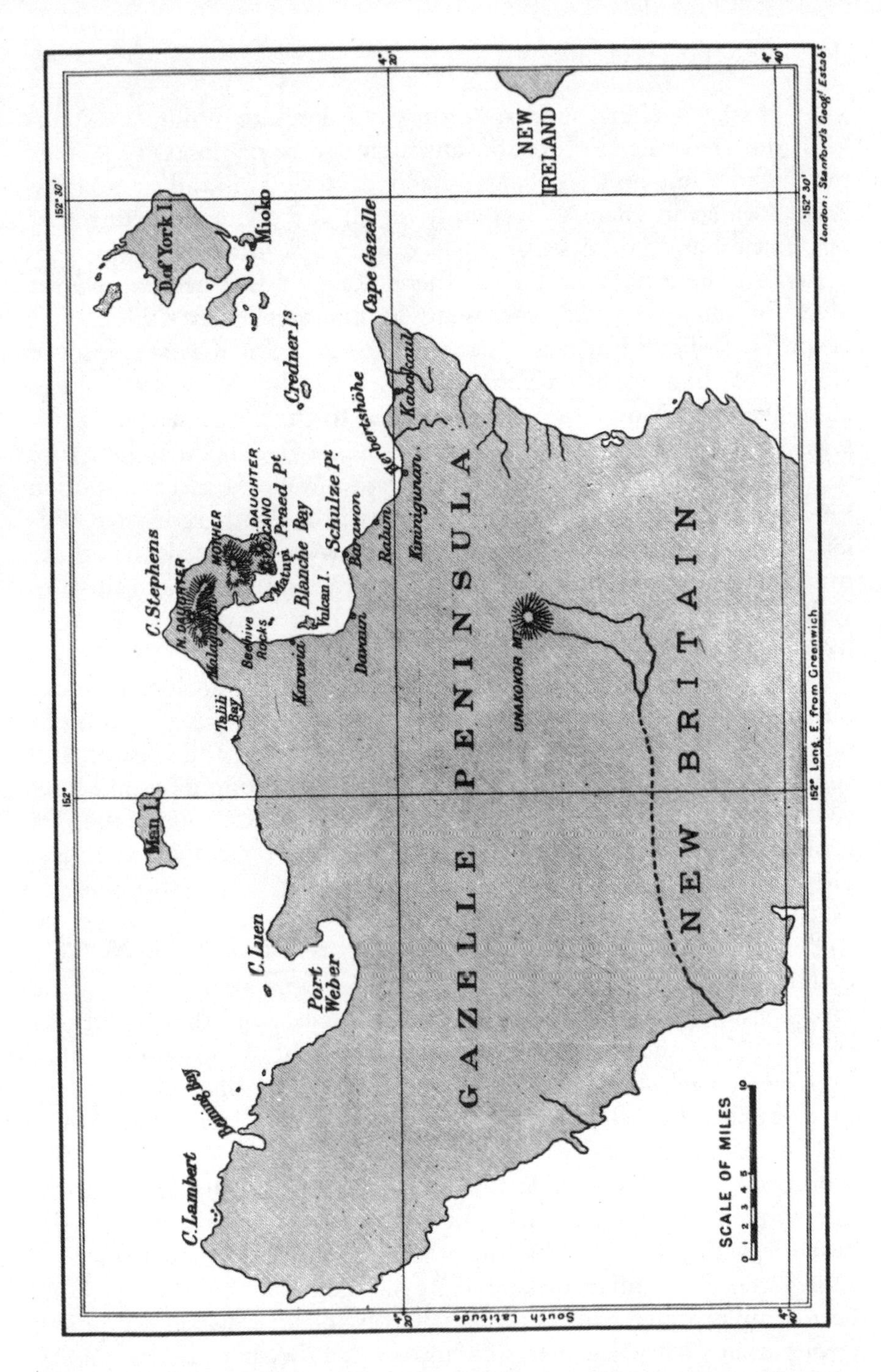

Map of the Gazelle Peninsula, site of Willey's major trapping efforts in New Britain.

If Father Helfer's sudden death was a shock to Willey, his lodgings must have been more so. Through choice or necessity, Willey purchased a hut on a small, uninhabited volcanic island named Rakaiya located in Blanche Bay near his site of operations. Purchase price: one hundred sticks of tobacco. Principal drawback: no fresh water. At the time Willey lived there, Rakaiya had been above sea level for only 16 years, following its emergence after a volcanic eruption. Willey's hut was apparently not prized real estate either; it was, according to his account, "loosely built" of sticks, with a palm-thatched roof. One wall was open to the elements (at least it was on the leeward side, Willey rationalized). The thatched roof was in poor repair, made poorer because Willey's assistants often kindled the evening fire with roof material they gathered when Willey was busy elsewhere. On top of that, Willey arrived during the monsoon season of continuous wind and rain squalls, when the prevailing wind shifts from the southeast to the northwest, so he must have spent a rather miserable time.

One of the shocks of the tropics for those who live in more northerly or southerly latitudes is how quickly the equatorial night falls. We of the earth's midlatitudes tend to associate the warmth and sun of the tropics with summer, a time for us of long days and long twilights. At the equator, however, the sun falls into the sea promptly at six; twilight lasts at most 30 minutes between the last glimpse of the sun and stygian darkness.

Willey lasted a month there before he sought new lodgings. But in other respects he quickly came to terms with life in New Britain. He realized that he would be far more comfortable if he made certain "accommodations" with the locals and their habits, as evidenced by the following passage from the 1902 monograph: "While [I was] staying at various places about Blanche Bay my food, besides tinned stuffs and an occasional fish, consisted principally of yams in the northwest season and taro during the south-east monsoon. These invaluable tubers and corms are purchased from the women at the markets which they hold periodically on the beach." With the wealth of fish in the region, it seems curious that Willey relied so much on starch in his diet. But it may not have been simple preference. Although he obviously relished yams and taro, perhaps he maintained this high-starch diet because his £2,000 grant did not translate into the all-important *diwara,* or shell money, the local currency at that time. Apparently during his visit he was a poor man as far as *diwara* was concerned. He makes one oblique reference to his lack of wealth: "By attending a funeral on one occasion," he wrote, "and demanding compensation for a stolen fish-basket on another, I came into a little property (diwara) but nothing worthy of mention." With regard to his diet, Willey noted: "As a general rule fish caught in the fish-baskets can only be bought with diwara, while yams and taro may be paid for with tobacco." Red meat and potatoes, those staples of the British Empire, were unavailable to Willey for most of his three-year voyage.

Willey's enthusiasm for taro may have been dampened a bit by the following episode:

> I accidentally incapacitated myself for serious work by biting into an uncooked taro. I had anchored on the reef at Nanuk [New Britain] for the night, it being high water, and intended to sleep on the boat while my boys lighted a fire and made themselves comfortable on the beach, cooking taro for themselves and for me. During the night it rained and thundered, and thinking to console myself with a taro I groped in the dark amongst the litter of ropes and gear, picked up one of these excellent fruits of the earth and made a lusty bite into it. It proved to be uncooked but I recovered from the dire effects within twenty-four hours.

Willey's relationship with the islanders appears to have been a blend of European racism, benevolence, and compassion. He was appalled by the physical condition of the local population, among whom serious staphylococcus infections were a common affliction that often caused disfiguring scars and other health problems. In Willey's day, before antibiotics were known, these staph infections must have been far more prevalent and dangerous.*

During his stay on the waterless volcanic island of Rakaiya, Willey attempted to provide medical treatment where he could, with unexpected results:

*Even today with plentiful supplies of antibiotics, staph infections are extremely common among many inhabitants of the equatorial Pacific. These infections generally begin as boils on the lower legs or arms that if left untreated can fester for weeks and become systemic, leading to general fatigue and debilitation. It is rare to find a Melanesian of the New Guinea region without terrible scars from these infections.

> The natives are born traders in their own way, and liberality is likely to have a demoralizing effect on them. But their ways are peculiar, and appear at times unreasonable; in fact their actions are often quite devoid of reason, being based either upon tradition or inherited instinct or else some chronic deep-seated, primary logical fallacy to which they are held in ineradicable subjection. Unfortunately besides these primal, more or less picturesque attributes, they well understand the subtleties of double dealing and sophistry. If, for example, one applies soothing balsam to their sores, they are ready to assume that the obligation rests with the donor, unless, perhaps, their idea is that a little tobacco is necessary to complete the cure.

Willey also observed local medicinal practices:

> For a time I placed much reliance upon the services of a man named To-mangiau, who was, indeed, something of a rascal, a *diable boiteux,* one leg being shorter than the other, but not without his points and a good swimmer. Shortly after I made his acquaintance he let himself be tattooed with a broken beer-bottle: two concentric circles over each breast blackened with burnt cocoa nut. I have also seen the natives using chips of glass and fragments of obsidian as lancets for blood-letting. The practice of bleeding at the seat of pain is resorted to in cases of headache, abdominal and muscular pains.

On another occasion, Willey witnessed local pediatric practices:

> All the infants at a certain age have their small bodies beset with numerous raised sores nearly an inch in diameter. To the inexperienced eye they present a shocking appearance at this stage, but one soon learns that the mothers cherish these sores and even, I believe, keep them open. If the sores do not break out there is a cause for anxiety for the future health of the child.

Willey also reports about fashion for women in New Guinea, circa 1896:

> Tattooing is not practiced at Panaete, but the teeth of the women are blackened and the ears pierced upon arriving at puberty. Should a young girl join the [Christian] mission, the old women of her village will urge her to leave the white man, saying, "Your teeth are not blackened and your ears are not pierced, no man will marry you."

TRAPPING IN NEW BRITAIN

Willey's primary mission was to determine the mode of reproduction of nautilus by collecting a series of embryos from various developmental stages. To do this he needed first to acquire the nautiluses, then induce them to reproduce. Nautiluses were well known to the New Britain islanders, who captured them in baited fish traps that were attached by hempen line to surface buoys. These traps were dropped 200 to 500 feet deep into Blanche Bay and then laboriously hauled to the surface by hand some hours later. The traps, constructed of wood and eight feet in diameter, were taken to the fishing grounds in open dugout canoes. Such fishing practices required reasonable weather conditions.

Willey and later workers soon discovered that acquiring nautilus specimens during the monsoon season was no easy matter. First it was necessary to procure an adequate supply of fresh fish with which to bait the traps. At the time of Willey's visit (or perhaps simply to impress this curious white man), the local islanders obtained the bait for Willey's nautilus traps by throwing sticks of dynamite into the sea and then scooping up the masses of shocked and dead fish that floated up to the surface after the explosion.

Willey described his first efforts at acquiring nautiluses as follows:

> On a typical occasion in January 1895, which I may describe, having procured my bait I took it to Davaun, To-vungia's village, in order to make arrangements for the setting of the traps during the coming night. I also purchased for myself a fish-basket in exchange for a "lava-lava" [a form of skirt worn by the islanders across the Pacific—hint that the adaptable Willey succumbed to clothing more practical than his Victorian costume] and six sticks of tobacco, and then returned to Vulcan Island. As soon as the fishermen who happened to be there saw that I was becoming independent of them they set to work with a will to bait the trap by tying the small fish called "malabur" on fibers purposely suspended inside from the framework. After this had been done and a heavy stone had been attached at each end, the fish trap was ready to be mounted upon a canoe and taken to the selected spot. We baited two traps on this occasion and started out, in two canoes, shortly after sundown, paddling towards Davaun and stop-

> ping rather close to the shore. We began to lower my basket at 7:20 P.M., using a native rattan cable, and finished paying out the rope after the lapse of twenty minutes; during this time, that is to say while the basket was slowly sinking, the canoe was kept in gentle and silent motion by the assistants. To-kiap, master of the canoe, having made a float of light wood, set the whole thing adrift and then we cruised around until 10:00 o'clock when we recovered the float and began to haul up. There were no *Nautili* in the basket, a disappointment of small moment to the biologist accustomed to negative results. We then moved a short distance until we arrived opposite "house belong To-gogi"; here we lowered once more, hauling up again shortly after midnight. This time we were fortunate in the capture of two specimens, a large mature male and an immature female.

These appear to have been the first of Willey's nautiluses. One wonders what Willey's emotions must have been when he finally held and viewed the objects of his laborious trip to the far end of the world.

Because of the incessant squalls during the early part of his trip, Willey must have been exposed to considerable risk, especially during the moments when traps were hauled from the bottom of the bay. He describes one night's work during this period when weather conditions could have proven fatal:

> One more incident may suffice to complete my description of the nautilus fishery in Blanche Bay. On January 21 [1895], four canoes, each carrying a baited fish trap, left Vulcan Island at 5:30 P.M. to sink baskets on the nautilus ground. I accompanied them as before in To-kiap's canoe. Having cast off the floats we went ashore at Davaun to rest, lying down on plaited cocoa nut leaves placed on the ground. At about 10:00 o'clock, a gale of wind and rain burst upon us and I adjourned, with To-mangiau and To-kiap, into the latter's house, a good weather-proof palm leaf hut, but very small. There were already two men asleep on the ground and a fire burning in the middle. However, we went in and lay down to sleep through the gale which lasted until midnight, and it was not before 1:30 A.M. that our members were sufficiently roused to set out for the purpose of raising our baskets. My basket contained six *Nautili,* and altogether the catch amounted to twenty-one, of which sixteen were males. Upon [our] commencing to return to Vulcan Island, the clouds looked so black and threatening ahead, lowering ominously over the summits of The Mother and Daughters [the active volcanoes overlooking Willey's trapping sites]

> that we reluctantly deemed it necessary to put back to Davaun. Accordingly we returned to enjoy the shelter of To-kiap's roof once more and hardly had we regained it when the storm clouds broke and converted the bay, for the time being, into a howling wilderness. We had in fact just been able to haul in our traps in the interval between two severe squalls. Just as the local inhabitants regard Blanche Bay as the source of the sea, so as I lay prone upon my mat listening to the raging of the elements it seemed to my fancy to be the very cauldron in which the northwest squalls were brewed.*

Even as well-intended and strenuous as Willey's efforts were to unlock the secrets of nautilus reproduction, they were hampered by his inability to keep specimens alive for any length of time after capture. The plan had been to keep mated pairs in aquariums, where they could lay eggs, but Willey found that nautiluses rapidly die in water temperatures above 80° F. And because seawater temperatures at the surface of Blanche Bay were higher than that all year, he was never able to keep his nautiluses alive in aquariums for more than a few hours. His alternatives were either to acquire nautilus eggs from the bottom of the sea or devise some way of putting freshly captured nautiluses back into the sea—perhaps in cages—and hope that during such an imprisonment they would breed. Willey actively searched for eggs and juvenile nautiluses by trawling the bottom of Blanche Bay, but without success. He also trawled the surface of the bay, hoping that young nautiluses were spending the early periods of their lives among the plankton. This effort, too, was unsuccessful.

Willey's memoir is full of ingenious plans for fulfilling his goals. This is how he describes his efforts to acquire nautilus eggs from Blanche Bay:

> Among my various devices designed to secure the eggs and young of nautilus, I employed a tangle-bar, a bar of iron with long hempen tangles hanging from it, which was sunk in deep water in places frequented by nautilus and left for a varying number of days with a buoy at the surface. I obtained the spawn of other cephalopods by this means, but not that of nautilus. Another obvious method of dealing

*It is this passage that makes me identify most with Willey and his voyage, for I have spent many nights both ashore and afloat listening to the raging winds of the so-called "Pacific" Ocean, and have cursed that mighty ocean for its efforts to thwart my own search for nautilus.

> with this intractable creature was to bait and sink fish-baskets in the usual way, and leave them down for several days instead of raising them at the customary time. Nautilus duly entered the traps, but in spite of sanguine hopes the experiment did not succeed, owing probably in part at least to various hindrances, such as the fragile construction of the cages, the inroads of conger eels and the buffetings of large fishes in the depths. A further hostile element which has to be reckoned with when operating in deep water is the periodical visitation of gales of wind, or half gales, especially during the northeast season. These will often mercilessly dispel all one's hopes and contrivances.

From January to September 1895 Willey was in constant pursuit of new specimens (he captured a total of 191 during that period) and conducted experiments and observations with holding traps such as those described above. But he was still quite far from his major goal: acquiring the embryological suite. As his stay in New Britain wore on, Willey apparently became increasingly restless and began to take ever-longer trips around the northern end of the island. In May 1895 he purchased a large, cutter-rigged sailboat. With a crew of one or two islanders, he explored and trapped from many areas of the Gazelle Peninsula of New Britain. We have no record that he had sailing experience or skills prior to his arrival in New Britain, but he or someone with him must have been a good sailor because his voyages were prodigious. His various study areas around New Britain yielded numerous nautilus specimens that became the source of important anatomical observations. They never gave him the eggs he needed, however, and by September Arthur Willey decided to leave New Britain for other locales in New Guinea.

EXPLORING IN NEW GUINEA

In October 1895 Willey obtained passage on an interisland schooner bound for Milne Bay, on the southeast end of New Guinea. After arriving there he was able to purchase another sailing cutter, the *Mizpah,* a vessel that was large enough for him to live on and use as a research station while pursuing more nautilus specimens. For the next four months he and "a crew of three mop-headed youths" explored the waters and islands around this end of New Guinea. Here he acquired the first complete specimen of *Nautilus scrobiculatus,* a species previously known only from its shell.

Willey traveled widely, visiting some of the world's most remote islands on what in a small sailboat must have been an epic voyage. His sailing ability was apparently put to test several times: "After much labor against the tide we managed to get a temporary anchorage off the westerly islet of the Conflict Group, intending to make the lagoon before dark, but failing to find the passage we spent a squally night tacking about." This rather matter-of-fact passage sums up what was probably a very uncomfortable and dangerous night. The island group Willey described is surrounded by extensive coral reefs. Hitting a reef in a wooden boat such as his during a squall would undoubtedly have wrecked the vessel; Willey and his crew must have spent the night searching for any glimpse or sound of waves breaking over the treacherous reefs.

Willey goes on to describe what happened the next morning, after tacking the boat all night:

> I had quite lost my bearings and suddenly found myself heading straight for the reef, but I got the jib and sail down just in time to avoid collision. It was not until the following sundown that we found the passage and anchored in the lagoon. Just about this time the southwest wind began to blow and I found the lagoon with its loose sandy bottom a most uncomfortable anchorage.

Willey and his crew were trapped in this lagoon by the high winds for five days.

> At the time of my visit the southwest squalls commenced, and [as the storm was] increasing in severity, my two anchors persisted in dragging, so that on the fifth day there was no other course open to us than to get clear of the lagoon, which had become a veritable trap, if we wished to avoid the ignominious piling up of our craft on the reef. After being foiled in several attempts we emerged successfully through a narrow passage nearly opposite to the one by which we had entered the lagoon. I think it must have been blowing quite half a gale and the sea was running high, but the *Mizpah*'s thin boards held together and we spent the night tacking and drifting in the usual manner.

After four months in the vicinity of Milne Bay, Willey was convinced that he was no nearer his goal than before and resolved to leave the region of New Guinea altogether in March 1896, some 16 months after his arrival from England: "By this time I had become convinced that it would be useless to spend more time in searching the coasts and islands of New Guinea for nautilus, in spite of the frequency with which the shells are cast upon the reefs. I therefore decided to go still farther east, either to New Caledonia, Fiji, or the New Hebrides, in any case by way of Sydney."

Willey rarely refers to his health during his time in New Britain and New Guinea. It is probable that he contracted malaria, however; this region of the Pacific is rife with that disease and with dengue fever, both of which are transmitted by mosquitoes. Willey makes oblique references to recurring fevers prior to his departure from New Guinea in March 1896: "I spent my last somewhat fever-stricken fortnight in New Guinea under the hospitable roof of the Rev. C. W. Abel, of the London Missionary Society, on the small island of Kwato opposite to Rogeia, within easy distance of Samarai, and left on board the SS *Titus* on March 28th, bound for Sydney by way of the Solomon Islands, the full enjoyment of this interesting voyage being marred by intermittent attacks of fever."

Willey traveled to Australia by steamboat, passing through the Solomon Islands. The voyage took three weeks, and Willey arrived in Sydney on April 20, 1896. There he apparently took a well-deserved rest. Once again we are left in the dark about this enigmatic man; it would be wonderful to know what trappings of civilization he luxuriated in after so many months of living in primitive, hot, disease-prone areas of the equatorial Pacific. He stayed in Sydney for over two months.

We do know that while in Sydney, Willey apparently spent some time at the Australian Museum, and it is there that he heard of another species of Nautilus perhaps worthy of investigation: *Nautilus macromphalus,* which lived around New Caledonia and the neighboring Loyalty Islands, some 700 miles to the east of Australia. With new resolve he decided to resume his quest for the embryological sequence on these islands. He left Sydney early in July 1896 on a four-day voyage by steamer to New Caledonia, then as now a colony of France.

After stops in Nouméa, the largest town on New Caledonia, and on the small island of Amedée, Willey sailed to the Isle of Pines, one of the most beautiful spots in the Pacific. Composed of sparkling white beaches and upraised limestone terraces, the island is covered with tall, graceful arucaria pines as well as the traditional tropical flora Willey was familiar with in the New Guinea region. After the humid tropics of New Britain, he must have found the climate on the Isle of Pines delightful; arriving in midwinter (New Caledonia is located at a latitude of 20–23° S), Willey probably never experienced more agreeable winter weather—temperatures in the mid-seventies and a steady, cooling trade wind. But sadly, he had no success trapping nautilus there.

THE LOYALTY ISLANDS

After two weeks Willey took the advice of an islander who had befriended him and decided to try his luck on one of the Loyalty Islands, a day's sail to the east by hired cutter. The first island he visited was Maré, where his stay was anything but a success. Willey had always been tolerant of and accepted by the various peoples of the Pacific, but he met his match in the French population on Maré.

"There is some difficulty in landing here," he wrote,

> as the rock-bound coast drops sheer into the sea, and the rudeness of the rocks simply anticipates the astonishing inhospitality of the natives. It is true I presented a dilapidated appearance, especially after walking over the scraggy limestone surface of the island, and had it not been for the good offices of a petty chief called Wainani, who had been a fellow passenger on the cutter, I should have been in a still more desolate condition, being already mistaken for a *libéré* [an escaped prisoner from Isle of Pines, at that time a penal colony]. The Government representative on Maré was the Commander, M. Journot, who entertained me at luncheon on my way to the great chief Neisselin at Netchi. With the best will, my friend Wainani none the less led me into a trap, by which I became exposed to the merciless and humiliating diatribes of Neisselin's wife, delivered in excellent English.

Willey's misadventures on Maré became increasingly comic at this point. One night he and Ito Pupu, the Maré islander he was lodging with, said their prayers and went to bed. In the morning, for reasons unknown, Ito Pupu confessed that his father had been a cannibal and had even eaten white men! Cannibalism was common throughout the nineteenth century in many parts of Melanesia, so this revelation could not have been too great a shock to Willey, but he did decide to leave Maré, and quickly.

After several more misadventures (including the capsizing of one vessel and the subsequent hiring of a second), and probably in despair after a month of fruitless endeavor in the New Caledonia region, Willey reached the island of Lifou, another of the Loyalty Group.

Here Willey's luck finally began to change. The geography of the area and cooler water temperatures of the surrounding sea made it possible to hold and maintain nautilus specimens. For the first time, Willey could keep them alive and study their behavior for longer periods. This is how Willey described Lifou: "The wanderings which I have described in the foregoing pages were all directed to one end. I wished to find a locality where nautilus could be both captured and kept prisoner in a moderate depth of water, in four or five fathoms instead of forty or fifty. Here on the shore of Sandal Bay, Lifou, I had come upon the ideal ground for which I had been searching."

Willey was able to trap specimens of *Nautilus macromphalus* from the water very near Lifou. Unlike New Caledonia and the Isle of Pines, where a barrier reef far from shore encloses a large lagoon, Lifou is surrounded by deep water and offered good nautilus fishing quite near shore. Also, the island was small enough that a lee shore could always be found to protect against the incessant trade winds. Willey stayed on Lifou for eight months.

As the natives in New Britain had done, the Lifouans used traps made of wood and bark to catch nautilus and fish. Willey reports that a variety of bait was used, including coconut crabs, crushed sea urchins, lobsters, and prawns. Apparently he had no boat in Lifou and was taken to the capture areas by raft. He described trapping operations as follows:

> The Lifouan rafts are well constructed and seaworthy. They are worked by sculling with a long, flattened pole, which passes through a hole in the centerboard at the front of the raft, and are sufficiently wide and buoyant to bear the weight of two men and a large fish trap between them. Provided that equilibrium is maintained, a rather difficult matter in choppy seas, the worst accident to be feared is the snapping of the pole.

Because the water temperatures were cooler, Willey found that nautilus could be trapped off Lifou at far shallower depths than in New Britain:

> The position of the fish traps, which are sunk to depths varying from three to eighteen fathoms, is ascertained by dead reckoning, buoys not being employed. They are dimly seen at sixteen to eighteen fathoms [96 to 108 feet, attesting to the clearness of the water] on a calm day, and when it is desired to raise them they are skillfully secured by means of a wooden hook, which is lowered from the raft and passed through the meshes.

This was Willey's first evidence of the powerful control that temperature plays in the life and depth distribution of nautilus.

REPRODUCTION AT LAST

Willey very quickly captured nautiluses and began placing them in closed cages in Sandal Bay, still hoping for eggs. And on December 5, 1896, after only four months in Lifou, Willey recovered a newly laid nautilus egg and promptly rewarded his Lifouan assistant, Saono, with the small fortune of 25 francs. Here is how Willey described the find:

> Having observed that the newly captured nautili adhered with great force by means of their tentacles to the sides of the vessels in which they were placed, I fixed some boxes in the cages. They attached themselves to the boxes but made no other use of them, and one day a huge conger eel effected an entrance into the trap and brought serious dissention into the household. To avoid a recurrence of such a disaster we closed up the entrance with sacking. On the next morning there was a curious white object, looking at first sight like some part of the axial skeleton of a bony fish, adhering to the sacking. Upon removing it we found that it had been tightly fixed to the sacking, and very soon I realized that it was the first egg of nautilus which rewarded my gaze. After two years of anxious groping in the dark it may be imagined what a thrill passed through my marrow, destined to be quenched when I found during the course of the following weeks that all eggs I attempted to rear were infertile.

In that single sentence, we pass from crowning triumph to despair. Nevertheless Willey spent four more months on Lifou trying various ways of incubating nautilus eggs. "I had several of these incubators [cages]," he wrote, "and tried all methods, pairing off the nautili in some, associating them in companies in others, manufacturing dark recesses in sackcloth, all to no purpose. The eggs

Nautilus eggs in the Nouméa Aquarium. The eggs are oblong in shape and usually cemented against some hard substrate. They are about 1½ inches long and approximately ¾ inch in diameter.

were all infertile and often simply consisted of the empty capsules without any vitellus inside."

Willey's patience finally came to an end following a spell of bad weather: "Near the end of January, 1897, we had a visitation of wind, probably the tail of a hurricane, which played havoc with my baskets, and cost the lives of some sixteen nautili." Although Willey does not say so, this loss must have been a terrible blow. *Nautilus macromphalus* is not easily found in large numbers in the New Caledonia and Loyalty Islands region; during his entire eight-month stay on Lifou Willey captured only twenty-six specimens, and with this one storm he lost more than half of them. He stayed on through February, then left Lifou by schooner on March 8, 1897.

Once back in Sydney, Australia, Willey telegraphed Cambridge to ask if he should continue his efforts to obtain nautilus embryos. The Cambridge dons encouraged him to proceed, so he set out once again for New Britain, arriving at Blanche Bay, the site of his original researches in the Pacific, on June 16, 1897.

Using techniques developed in Lifou, he stocked cages with nautilus in Blanche Bay, this time at depths of 180 to 300 feet. He soon had eggs from a second species of nautilus, *N. pompilius,* to compare with the *N. macromphalus* eggs he had previously acquired off Lifou. These eggs were similar in appearance to the others and were also infertile. Willey was only the first of many investigators to acquire nautilus eggs, then wait and watch as passing weeks brought no evidence of development. Never in his lifetime would Willey see the object of his three-year quest.

Willey stayed in New Britain for three more months, and in September 1897, almost three years to the day after he had set out from England, he left the islands to return home.

RETURN TO ENGLAND

After arriving in England Willey joined the faculty of Guys Hospital of London, where he became a lecturer in biology. He stayed in this position for two years, and during this period completed his monograph on the anatomy of nautilus, including the description of his voyage.

Willey's monograph is a masterwork. It shows that he knew more about nautilus than anyone living, and certainly understood its anatomy better than anyone before him had. In some respects Willey did his job too well. For the most part his main achievement —his exacting descriptions of nautilus anatomy—tends to diverge from the scientific process, because they are simply data gathered in the vacuum of testable hypotheses. His observations are so detailed and documented that as descriptions of gross anatomy, they are a job finished. One should not sell Willey short, however; in his monograph he did discuss which anatomical characters are homologous or analogous to structures of other mollusks and, based on these findings, suggested the degree of similarity among various molluscan groups. His approach also allowed him to make shrewd guesses about the course of embryological development that occurs in the fertile eggs he was never able to acquire. However, it was not Willey's detailed anatomical descriptions of *Nautilus pompilius* and

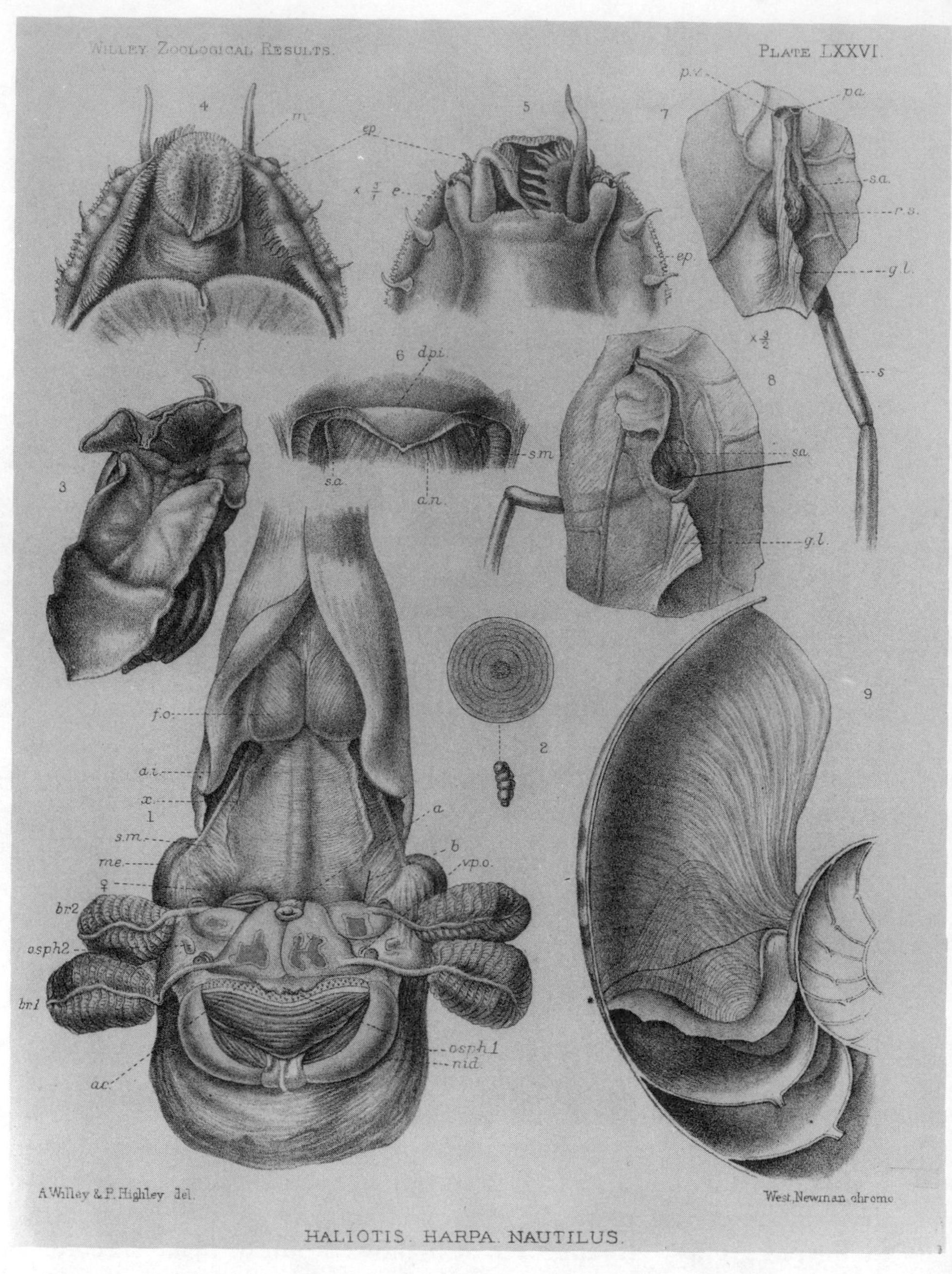

Cutaway view of nautilus shell and underside of soft parts. The spherical object at center is an enlargement of a nautilus kidney stone. These kidney stones are thought to be associated with calcification of the shell.

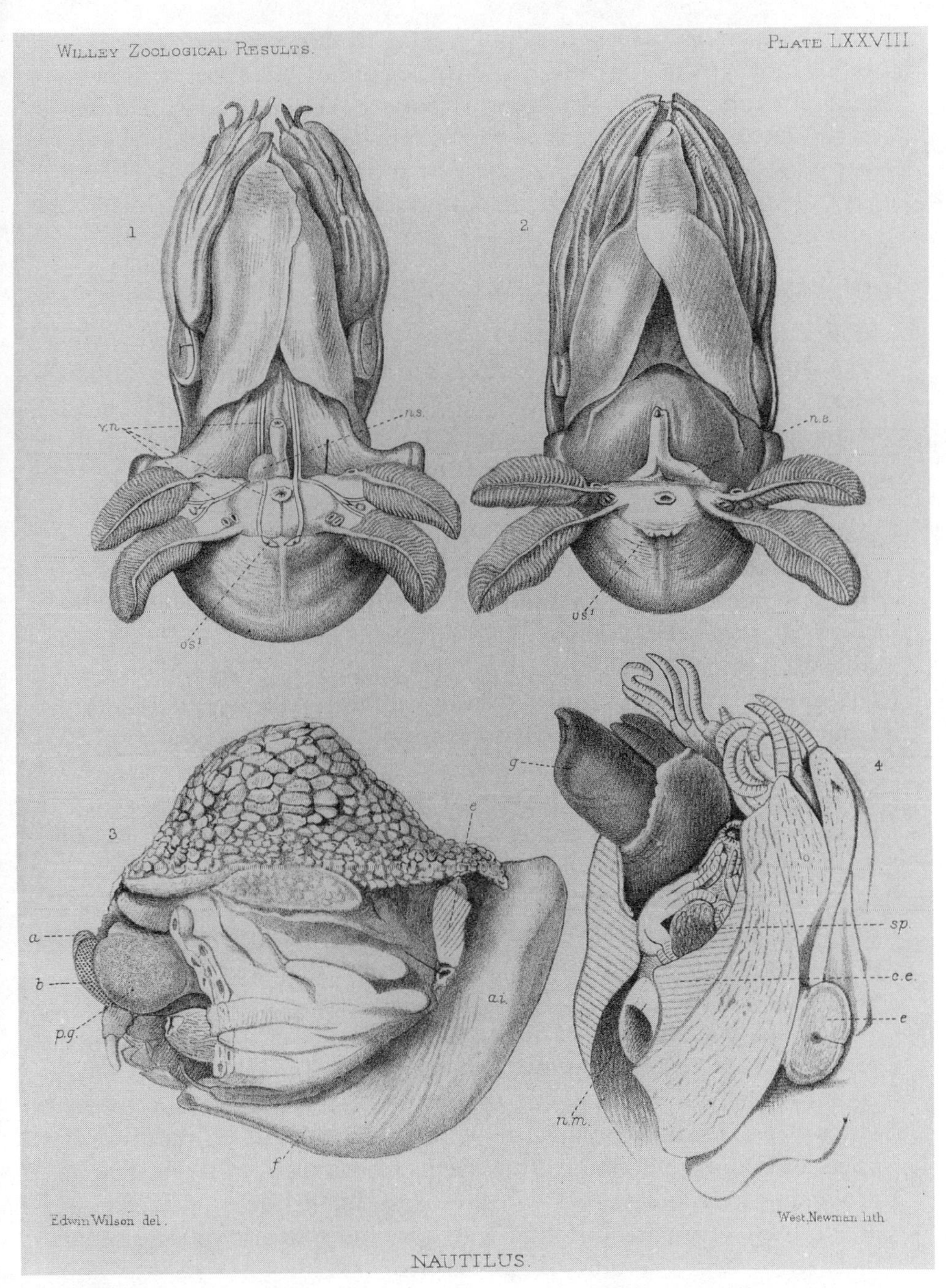

The underside of a nautilus, as well as the first illustraion of the soft parts of the rare species *Nautilus scrobiculatus*.

Nautilus macromphalus that inspired other young scientists to follow in his footsteps in search of nautilus, but rather, his short comments and notes about buoyancy control, growth, ecology, and the differences among the species that stimulated much further interest and became the sources of hypotheses for future inquiry. Willey's findings on each of these aspects are summarized below.

BUOYANCY AND GROWTH

Willey made his most grievous errors with regard to the way nautilus forms and maintains buoyancy. Since the time of Robert Hooke in the seventeenth century, it was thought that the air chambers that give the nautilus its buoyancy were created in the following way: The nautilus shell may be considered a long tube wound around itself; as the animal grows, it forms new shell material at the aperture, thus making the tube larger. The soft parts of the nautilus (the tentacles, head, and viscera) were believed to sit in the most recently formed one-third of this tube; the rest of the tube constitutes the air chambers. As the nautilus grows, it must periodically move its soft parts forward in the shell and lay down new chambers in the spaces most recently vacated. Willey's mentor, Ray Lankester, had proposed a theory known as the "preseptal gas hypothesis." According to Lankester's idea, the back of the nautilus's body periodically moved away from the last-formed septum, which it had rested against. As this posterior body section moved forward in the shell (toward the shell's aperture), gas was secreted into the space left behind. When the back of the body had moved far enough forward, a calcareous substance was secreted to create a new partition, or septum, thus forming a new chamber that was already filled with gas.

Willey was the first scientist in a position to check this idea and seems to have accepted it without question. Regarding the buoyancy process, he wrote: "The buoyancy of the shell is due to the series of air chambers which have long excited the admiration of poets and philosophers. Nautilus seems to have a peculiar faculty of producing gas, and my knowledge of the pallial veins was chiefly due to their automatic injection with gas after removal of the animal from the shell." Willey supposed that the gas went into the chambers through the veins located along the posterior mantle, a thin sheet of tissue lining the back of the nautilus's body. It is this organ that secretes the septa. Willey conjectured that once the gas had been secreted and sealed within the chamber, it played no further role in buoyancy change: "It is, I am convinced, an error to suppose that variations of pressure of the air in the chambers enable nautilus to rise or sink as the case may be. The air renders the shell buoyant once and for all."

In this same section, Willey made a crucial observation that should have told him that the preseptal gas hypothesis could not be correct. He noted: "If the shell, with the live nautilus in it, be perforated over the chambers under water, the air bubbles gently out as the water enters." This passage suggests that gas pressure within the chambers Willey examined must have been about one atmosphere. Had the pressure been higher, Willey would have noticed a loud hiss instead of gentle bubbling; had it been lower, water would have entered rapidly before any gas escaped. If we match these observations with the principles of the preseptal gas hypothesis, we arrive at the following conclusions: 1. Nautilus builds new chambers by moving forward in its shell; the back of the animal moves away from the last-formed septum. 2. This vacated space is filled with gas produced by tissue or veins at the back of the animal's body. 3. When the space is large enough, it is sealed off with a calcareous septum secreted from the back of the body. 4. Once the gas is inside, the nautilus cannot regulate its pressure. 5. The gas is approximately one atmosphere in pressure.

These requirements seemed reasonable to Willey after he had studied more than 200 nautiluses. However, many of those he observed had been captured in deep water, some from as deep as 500 feet, where ambient pressure is about 250 pounds to the square inch. The main problem with the preseptal gas hypothesis is related to the high ambient pressures that exist at the great depths where nautilus normally lives. How would the preseptal mechanism of chamber formation have to work at such depths? The gas pressure that would allow the nautilus to form any space—any volume—between its liquid-filled soft parts and its shell must equal *at least* ambient pressure. For a nautilus at a depth of 500 feet, this would be about 250 pounds to the square inch, which is very high pressure indeed for any biological organ to produce. A teleost fish can accomplish this feat in its swim bladder, but this is a very specialized organ system and produces the gas in small volumes. According to Lankester and Willey's idea of how new septa form, high-pressured gas in nautilus is produced by simple, undifferentiated veins of the posterior mantle, the soft parts that secrete and later rest against the septa.

Boyle's Law describes the relationship between pressure and gas volume. Unless gas were released somehow during ascent, gas pressure in the chambers would *increase* as the nautilus rose to the surface. If a nautilus found at a depth of 500 feet, with its initial gas pressure of 250 pounds per square inch, were brought to the surface, the pressure within each of its chambers would increase by a factor of approximately 17. This means that the gas pressure within the chambers would be about 4,250 pounds per square inch by the time the nautilus reached the surface of the sea. This is four times more than the pressure of about 1,100 pounds per square inch at which the nautilus shell is known to implode or explode.

Had Willey found such a pressure upon opening a nautilus shell, he would not have described it as a "gentle bubbling of gas." It is surprising that neither Lankester nor Willey saw the problems in their hypothesis of buoyancy maintenance and new chamber formation, especially in light of the known depth ranges of nautilus.

Even more surprising is that Willey never observed liquid in the shell of a nautilus. As we will see in the next chapters, when a nautilus forms new chambers, the space between the advancing body and the last-formed septum is filled with *liquid,* not gas. This liquid must completely fill the new chamber; it is later removed and replaced with gas at very low pressure. Some liquid remains, however; on close examination, every chamber contains a small amount. Newly formed chambers may be either partially or completely filled with liquid, depending on their age. In a large nautilus, as much as half a cup of saline liquid may be found in the last-formed chamber. In immature, growing nautiluses, the liquid within a partially completed chamber is immediately visible when the soft parts of the body are removed from the shell. It is incredible that Willey, who cut open so many shells, missed this entirely.

Although he describes in detail the morphology of newly secreted septa, he never mentions the liquid that completely fills each new space they delimit. "The foundation of a new septum," he wrote, "consists of a very thin, easily torn membrane. . . . When a calcified septum is substituted for the primary membranous septum by the deposition of nacre, it is at first exceedingly thin and fragile, becoming gradually thickened by further secretion, but the size of the air chamber is determined from the commencement of calcification and undergoes no subsequent change."

Why Willey does not mention the large volume of liquid behind the membrane and calcifying septum is anyone's guess. In virtually everything else, Willey's observations have been confirmed, and we know that he was a careful and patient observer of details. It may be that his crucial first observations of what we now call cameral liquid were omitted from the final monograph by the external refereeing system. Ray Lankester, director of the British Museum and the world's reigning expert on nautilus, certainly must have at least seen the manuscript, and may even have edited it. The observation of cameral liquid would have been a mortal blow to his widely quoted preseptal gas space hypothesis. Sadly for science, and for several generations of invertebrate zoologists and paleontologists, this erroneous hypothesis rested unchallenged for an additional sixty years.

Another item Willey failed to make reference to in his monograph was the growth rates of nautilus. Although he describes in detail the changes found in the shell as the animal reaches sexual maturity, he makes no guess about the amount of time that might elapse between hatching and the point at which a nautilus attains its determinate full size and ceases growing. Nor did he speculate about how long a nautilus might live after reaching maturity. (These aspects of growth had to await the latter half of the twentieth century.) Willey did note a discontinuity on the early part of the nautilus shell that perhaps related to hatching from the egg; this mark, a small groove on the shell's exterior called the nepionic constriction, suggested to him that nautilus hatches with seven chambers already completed.

ECOLOGY

During his three years in the Pacific Willey made several interesting observations about the mode of life and ecology of nautilus.

Because at the time there was no equipment available for diving, Willey was unable to make direct observations about the behavior of nautilus in its natural habitat. Many of his observations about distribution and mode of life are therefore based on inference or on information available from specimens captured in baited traps. With regard to nautilus's diet, Willey quoted Richard Owen's earlier observation that crustacean fragments had been found in the animal's crop region. Willey concluded that nautilus feeds off the bottom rather than in the upper reaches of the open sea: "It is also desirable to remember that nautilus obviously draws its supplies of food from the bottom of the sea, it is a ground feeder. . . ." His descriptions of various means of trapping nautilus specimens suggested that crustacean material was the most effective bait, and he concluded that prawns and crabs made up the creature's major food supply.

Although Willey's comments about food and feeding are interesting, by far his most controversial observation about the habits of nautilus concerned migration. He suggested that during the night nautilus migrates from its daytime deep-water habitats to shallower waters in search of food. Willey's original statement, published in 1899, is as follows: "In fact I came to the conclusion in New Britain, which I afterwards confirmed in the Loyalty Islands, that the feeding ground is not the breeding ground—in other words, that nautilus migrates in shoals nocturnally from deeper water into shallower water in quest of food." Here he says nothing about the frequency of these journeys (he never says that all nautiluses migrate every night) or about the horizontal or vertical (depth) distances of such migrations.

Willey amplified this statement in his 1902 monograph, noting that nautiluses are rarely found at the surface, and then usually only because they are sick or dying. He wrote: "When nautilus has been taken, as a great rarity, at the surface of the sea, it has generally, if not always, been found that the specimen was in a more or less moribund condition. At the same time, with its known faculty for swimming and migrating in some places into quite shallow water a few fathoms in depth, it is quite conceivable that an individual specimen might occasionally wander away from its home and arrive at the surface, but there is no evidence that this is a regular practice." His comment about finding the animals at a depth of a few fathoms clearly refers to Lifou and not New Britain, where surface water temperatures are far higher and thus make it necessary for nautilus to live deeper. About the animal's social habits, Willey noted in his 1902 monograph that "the evidence of the traps goes to show that nautilus is a gregarious animal and nocturnal in its habits. It repairs in shoals at night to its shrimping grounds, but I suspect that it breeds in deep water or in inaccessible submarine gullies."

These few sentences, extensively quoted and misquoted, were the only clues to the nautilus's mode of life for many years. Willey's work was so influential and knowledge about nautilus so scarce that interpretation of the behavior of many extinct nautiloid and ammonoid cephalopods long depended on his observations, especially that of nightly vertical migration.

The idea that nautilus moves upward from a deep daytime habitat to shallower waters at night also began to influence theories about its buoyancy system. Zoologists and paleontologists argued that such large-scale depth changes in a single night would require far more energy than swimming alone could reasonably provide. The nautilus buoyancy system began to be pictured as the force that permitted the hypothesized vertical movements. According to this perception, a nautilus would make itself positively buoyant at the start of the night and float up to shallow water, then reduce its buoyancy at the end of the night and sink back to the greater depths of its daytime residence. This idea became common in most descriptions of nautilus in zoology and paleontology texts and articles written between 1900 and 1975.

For instance, in an otherwise admirable paper about the potential for vertical movement in ammonites, the English paleontologist W. Heptonstall wrote in 1970: "It is known that nautilus spends the hours of daylight on or near the seabed at depths to 600 m and comes to the surface at night (Willey, 1899). In order to migrate vertically through such a distance in no more than a few hours, the shell must undergo significant changes in density." In fact, Willey never stated that he had observed that a 600-meter vertical migration had taken place in a few hours. As late as 1978 two Japanese scientists, Kawamoto and Mikami, rather severely paraphrase Willey's original comments and claim he "reported that the nautilus makes a diel [daily] migration from a depth of about 500 m up to the surface of the sea. However," they continue, "it is axiomatic that such a long-ranged migration within a short period of time cannot be managed by only water-jet propulsion through the hyponome."

The origins of Willey's idea about some sort of nocturnal upward migration are completely mysterious. Although he says he first came to his conclusions in New Britain, I believe that they were based instead on observations and sightings of nautilus in Lifou. Spearfishermen and divers there often encounter nautilus at a few meters' depth at night, but never during the day. It is quite possible that lobster fishermen in Lifou told Willey they had seen or even caught nautilus at night (lobsters are nocturnal and live on the same sides of the reefs that nautilus does). In any case, Willey's rather innocent remarks about nocturnal movement are those most often quoted and misquoted.

SPECIES AND RANGE

When Willey made his voyage to the Pacific, four different nautilus species had been defined. Each was distinguished by the morphology of its shell. *Nautilus pompilius,* the most widespread, ranges from Fiji west and north to New Hebrides, the Solomon Islands, the New Guinea region, and the Philippines. This was the species Willey captured and described from New Britain. He knew that the other species were far more restricted in range and generally lived near a single island or archipelago. These were *Nautilus macromphalus,* the species from which Willey first succeeded in getting eggs, known from New Caledonia and the Loyalty Islands; *Nautilus scrobiculatus* (which Willey knew by the name *N. umbilicatus*), found at Manus Island, New Guinea; and a fourth species, *Nautilus stenomphalus,* found near Australia's Great Barrier Reef. During Willey's time the latter two were known only by their shells, because the soft parts were unavailable to the nineteenth-century scientists who classified them as separate.

The two best-known species, *N. pompilius* and *N. macromphalus,* differ markedly in their shell morphology. In their soft anatomy, however, they were found to be indistinguishable. Willey was understandably pleased when he was given the first known soft parts of *N. scrobiculatus* while visiting Milne Bay, New Guinea; in this single specimen he saw that the soft parts as well as the shell differed from the other two species. "I was therefore very pleased to come into the possession of a single mutilated specimen of *N. umbilicatus* [*N. scrobiculatus*] accompanied by its shell, which had been picked up from the surface of the sea, not far from Milne Bay in British New Guinea, and to find that this species differed noticeably from its congeners by the character of the hood, the gibbosities of which have the form of a flat-topped angular area separated by deep grooves, producing a pronounced tesselated appearance." Another individual of this species would not be found for 90 years.

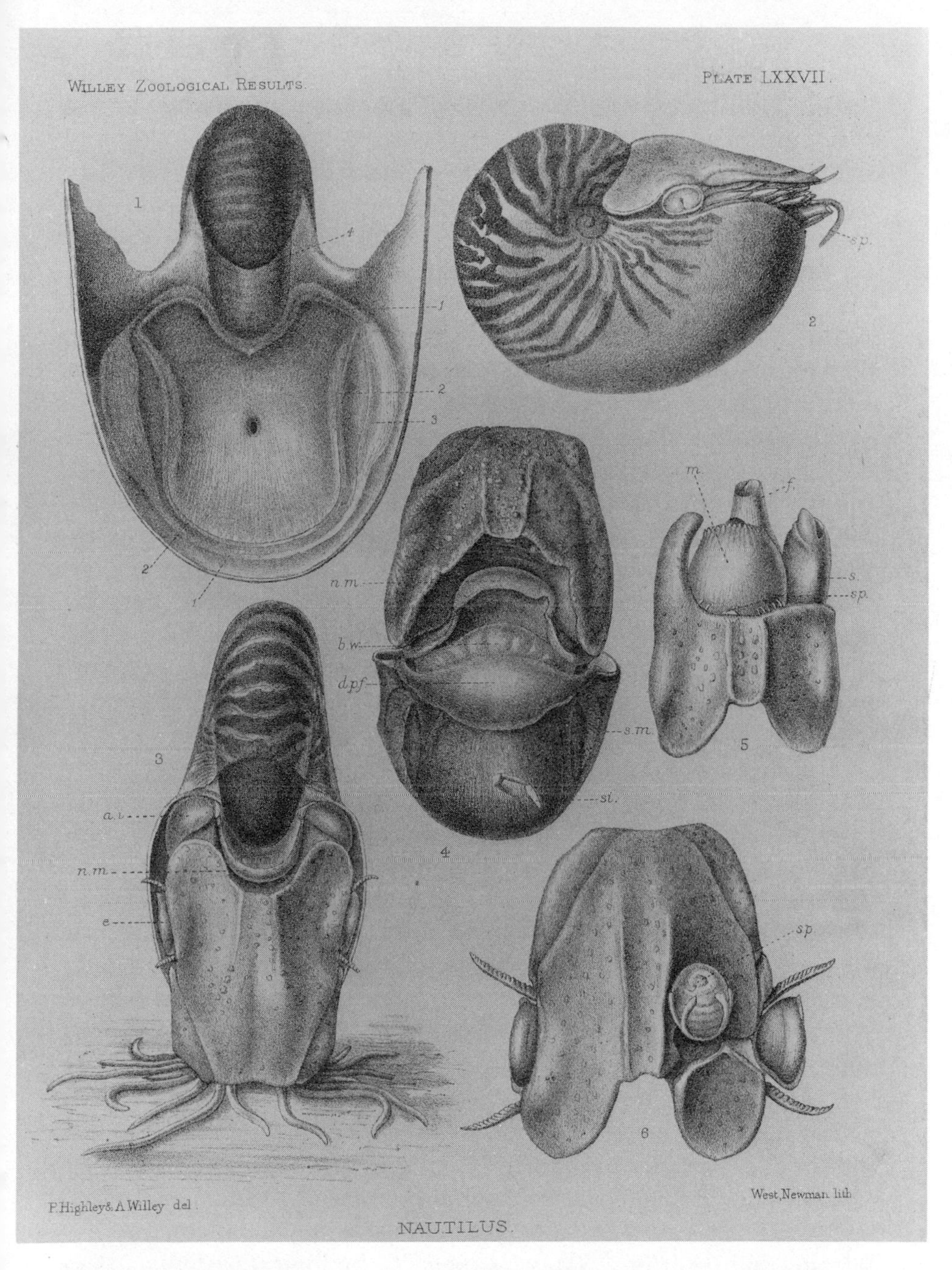

Nautilus macromphalus, the species Willey studied in the Loyalty Islands. This species is also found in New Caledonia.

Willey was very interested in the geographic ranges of the various species and in the biogeographic information these ranges suggested. He understood that nautilus was not a particularly good swimmer or well suited to traversing wide oceanic distances, and that its reproductive process was ill-suited to wide dispersal of young. Willey noted: "It is curious that so far as is known with any certainty it [nautilus] does not occur west of the Strait of Malacca nor east of Fiji. Although it descends to deep water, the single specimen obtained during the Challenger Expedition [the great nineteenth-century oceanographic voyage] having been dredged from 320 fathoms [1,920 feet], it is not an abyssal form, but rather seems to affect the vicinity of large islands, which perhaps in former ages were united to still larger continental masses. Charles Hedley, who has developed a remarkable theory of a Melanesian plateau that once united New Caledonia to New Zealand (as indicated primarily by the distribution of land mollusks), says in a recent paper: 'It is remarkable how strictly nautilus observes as its eastern limit the ancient coastline of the Melanesian Plateau.'" This passage suggests enormous prescience on the part of Hedley, who anticipated the plate tectonic theory, formulated in the 1960s, that would corroborate his land snail findings.

REPRODUCTION

Even though Willey failed in his main task—acquiring nautilus embryos—he made many new observations about breeding, egg laying, and the anatomical differences between males and females. Willey was the first to observe nautilus eggs, which for cephalopods (or any invertebrates, for that matter) are very large. The eggs of both *N. pompilius* and *N. macromphalus* are oblong and nearly an inch and a half long. Based on this egg size and on certain morphological features of small nautilus shells—the sculpture on the earlier part of the shell is decidedly different from that on the later part—Willey deduced that the nautilus would hatch with a fully formed shell of about an inch in diameter.

Another finding concerned the distribution of the sexes. In a sample of 216 nautiluses—*N. pompilius*—Willey captured in New Britain, the majority (150) were male. In biological systems an excess of males over females is very rare. Although Willey discovered the reverse to be true for the *N. macromphalus* he captured in Lifou, the low number of specimens he took there (26) makes this finding insignificant. Later researchers have generally duplicated Willey's results, trapping more males than females.

Willey also found a distinct sexual dimorphism—males are larger than females—and differences in the outline of the apertural region of the shell.

After Willey concluded his writings on nautilus, he left England for the other side of the world once again, this time for Ceylon (now Sri Lanka), where he took over the directorship of the Colombo Museum. The year 1902 was an eventful one for Willey. Not only did he begin his new job in Ceylon; he also married, his monograph was published, and he was made a Fellow of the Royal Society of England, the highest honor a scientist of the British realm can receive except a knighthood.

Willey stayed in Ceylon until 1910, when he accepted a professorship in zoology at McGill University in Montreal, Quebec, Canada. There he stayed until his death in 1942 at the age of 75.

Following publication of his 1902 monograph, Willey never returned to the subject of nautilus. A passage in the opening statement suggests why: "My efforts were only crowned with partial success, an eventuality for which I had prepared my mind beforehand." Similarly, he introduces the substantive part of the book by belittling his considerable accomplishment, saying, "This is neither the time nor the opportunity for the composition of a costly monograph on nautilus."

Willey's journey to study nautilus was epic. He endured sickness, endless hard work, and ultimate disappointment. But his work is a lasting monument, and his research in the areas of buoyancy, ecology, species, and aspects of reproduction provided the impetus for future voyagers in search of nautilus.

Chapter 4

NAUTILUS OBSERVED: ANNA BIDDER, 1960

Arthur Willey's monumental voyage and subsequent monograph electrified the scientific community, but as the world hurtled into World War I, the Depression, and World War II, nautilus was left to its lonely scavenging on the deep fore-reef slopes of the tropical Pacific Ocean. The only progress during this period came from a French physiologist named Purvot-Fol, who was able to examine the soft parts of nautilus and realized that Lankester's preseptal gas hypothesis was nonsense. Her documentation of this conclusion in 1937 was overlooked, however, and as late as 1966 the preseptal gas hypothesis was still being presented as fact in scientific sources.

The next great voyage to the Pacific for the study of nautilus did not occur until 1960. It was made by a determined English zoologist named Anna Bidder. In March 1986 I talked with Bidder in Cambridge, England, about her experiences and about nautilus, for she is the bridge between Willey and Cambridge past and the modern expeditions. When Anna Bidder went to the Pacific to study nautilus she was fifty-seven, and the path she followed was very much the same as Willey's.

During my stay in Cambridge a great storm blew out of the North Sea, and for two days wind and rain swept across the fens, uprooting huge oak trees and knocking down one of the huge sixteenth-century spires of King's Chapel College. Whenever the sun managed to break through the gray winter sky, I rode a decrepit bicycle along the narrow streets to visit Anna Bidder at 2 Cavendish Place, to be transported back in time and to the faraway Pacific.

Anna Bidder dissecting a giant squid, 1961.

Anna talked with me in the most beautiful English punctuated by hearty laughs. We drank tea, and occasionally she would pull lustily on a cigarette. I taped our conversations, and her comments and observations are transcribed here verbatim. She first took me back to Willey's Cambridge, when her father, George Bidder, was a famous professor of zoology there and she was a very young girl. Anna was born at the time Willey was leaving for Ceylon; while one searcher of nautilus was finishing forever, a future searcher was just starting out.

I asked Anna about her memories of Cambridge in her childhood, a Cambridge well known to Arthur Willey. She was full of stories.

> At the time of Willey, in the nineties, the Zoology Department was a small department, and Newton would have been Professor. Everything was so very much smaller. At that time, of course, no Fellow [a faculty member of one of the Cambridge colleges] could be married. They may not have all had to be in holy orders, but they certainly couldn't be married. What was absolutely *not done* was to ask a Fellow what he had done during the vacation, or what he was going to do during vacation. My father was friends with a man named Jackson, a classicist who was an old man, and my father was young. And my father told me that he had asked Jackson what he had done on vacation, and Jackson dropped on him like a ton of bricks because they all of them had perfectly respectable wives and families somewhere else. And if you did ask, they would say, "I should be going into the country." And as soon as this rule was changed and Fellows were allowed to be married, they all moved their families into Cambridge.
>
> When I was a little girl my father and mother were entertaining on Saturdays and Sundays, graduates and undergraduates, and my mother would go through the Residents List at the start of every term, looking for the sons of friends of theirs who had to be asked to lunch. I had to hand around cigarettes and play daughter of the house.

The circle of zoologists in England around the turn of the century was a small one. I asked Anna if her father had ever come into contact with Willey.

> My father knew Willey, of course. I gather that Willey was a nervy type. One is full of admiration for what Willey did when you realize that he was probably very highly strung and a bit soft, and then he went off and lived this very tough life and was very ill. He picked up a fever when he was staying on Vulcan Island in New Britain. Father, who was a friendly person, didn't speak of Willey as a friendly person. At least Willey had given Father a copy of his monograph, but it got pinched.

I asked Anna how she had become involved with nautilus.

I went into zoology because my father said that I should like it, and I did. I was sent out to Switzerland to begin life as a research student to study with Adolph Portmann [one of the greatest European zoologists]. Father very wisely wanted to get me out of Cambridge, where I had lived all of my life, and into another environment. At that time German was the second language of science, but at that time there was no department in Germany appropriate for me, with the right kind of work going on. Adolph Portmann was a very young university assistant on a princely salary of below 200 francs a month. He was of course doing everything he could in his spare time to stay alive, illustrating books and giving lectures and all the rest of it. The senior professor there had a whole list of research projects for me, mostly dealing with parasitic worms, but Portmann produced two of the things he had started when he was playing around at Roscoff [a French marine laboratory], and one of these dealt with the yolk absorption in some *Loligo* [squid] eggs.

I always had a tendency to mollusks. I'd liked mollusks partly because as a child I had liked pretty shells, and also because as a lazy student I didn't like crustacea, because they had too many things to count and remember by name—all these wee little bits of joint. A mollusk is a nice simple animal; the only thing you have to remember anything about is the radula, so I took to mollusks. But anyway, the free adaptation, the way the simple shape of a mollusk can get pulled out into so many, many, many different forms—that always fascinated me. But the cephalopods I hadn't known very much about or bothered with.

Newly hatched squid have a lot of internal yolk, which is not absorbed into the bloodstream, as it is in a chick, but apparently goes straight into the digestive system. This prompted us to ask, What is the digestive system doing in the adult? And so I started to look. The paper didn't get published until 1952, even though the research was done in the twenties. But there it was, an account of the digestive system of *Loligo,* and no one had known how that had worked before, and it was all rather fun.

In the thirties I started looking around at other cephalopods and found that there was a constant plan, on the part of the digestive system, running through all of the cephalopods I examined, whether they were octopuses or what. And then, of course, I wanted to know what the digestive system was like in nautilus, and I managed to borrow some of Willey's material from the British Museum. I found that nautilus had got this curious organ in the digestive system called the spiral caecum, just like all the others, but that it worked just the other way around; it was, so to speak, a left-handed coil instead of a right-handed coil as in all of the coleoids. And I found all sorts of elaborations in the nautilus tract itself that were quite different. It was fascinating!

When I looked at the existing accounts of nautilus, I found that the previous workers had not gotten it right, but had known that something interesting was there. Well then, of course, one longed to get at the animal. But at that time I was keeping house for my father after my mother died, and then came the war and the house was filled with evacuees, and then my father was getting old and there was no question of leaving him for long. My father finally died in 1953, and that left me free, when I was fifty.

At that time I said to Carl Pantin, a fellow staff member in the Cambridge Zoology Department, on the occasion of the departure of another staff member leaving for New Guinea on an entomological collecting trip, "Carl, I wish I could get him to get some nautilus material for me," and Carl said, "Go and get it yourself, Anna." I blinked and set to work and began to collect information from different people, and found out what it would cost, and got advice that *of course* I must fly, because any grant giver would think it altogether wrong if I tried to save money by going slowly, so I must put in for air transport.

In 1952 I attended an oceanographic conference attended by Jacques Piccard, the Frenchman who piloted the bathyscaph to great depths in the Mediterranean and the Pacific. I got an introduction to him, and he knew of a man named Dr. René Catala, director of a new aquarium in Nouméa, New Caledonia. Piccard got me an introduction to Catala, and I wrote to Catala, who replied that no nautilus were captured in the waters of New Caledonia, but that they could be caught off Lifou in the nearby Loyalty Islands. And then I began to seriously make plans and applied for a grant from the Nuffield Foundation, and got a grant for £2,000, which was indeed a generous grant for those days. My round-the-world ticket was £800, which gives you a sort of measure of how much money it was. Let me tell you that a round-the-world air ticket is quite a thing to have!

I got tips that whatever you do, take enough containers, so I had jam jars and slop pails and washing-up basins from Woolworth's, as well as some proper scientific plastic stuff. Plymouth [the Marine Biological Laboratory at Plymouth, England] was terribly helpful in getting me going. I was backed because Plymouth wanted someone to go out and do a pilot project to see if nautilus was readily obtainable and if it could be handled so the physiologists could go out and do proper work on it.

I set out in September 1960 for New Caledonia. By this time

> Catala had found out that nautilus could be captured on the barrier reef in front of the capital city of Nouméa, where the aquarium was located. I spent September to December 1960 studying and observing nautilus in Catala's aquarium in New Caledonia.

The New Caledonia aquarium Bidder visited had a magnificent running seawater system in which nautiluses could be kept at a water temperature cool enough to ensure their survival for months at a time. During her three-month stay, Bidder had 40 specimens of *Nautilus macromphalus* to observe at the aquarium. The results of her research there were published in 1962 by the British journal *Nature*. Although a very short paper, this communication represented the first significant scientific advance in the understanding of nautilus since Willey's 1902 monograph had appeared.

Bidder's paper in *Nature* was divided into three sections. In the first she described how nautilus uses its tentacles for feeding and receiving sensory information. In the second section she described, for the first time, its swimming and respiratory movements and how these change with various behavioral activities. The final section dealt with buoyancy control; here Bidder made the first report that the nautilus's chambers contained liquid as well as gas.

The discovery that the chambers of a nautilus contained salty water as well as gas was of fundamental importance. At the time of her Pacific voyage, Bidder knew of the recent (late 1950s) findings of two Plymouth physiologists, Drs. Eric Denton and John Gilpin-Brown, on the role of liquid in the formation of new chambers and buoyancy control in the cuttlefish *Sepia*. Denton and Gilpin-Brown, not convinced of the preseptal gas method of new chamber formation hypothesized by Lankester and Willey in the late nineteenth century, were certain that nautilus, like *Sepia*, would prove to have chamber liquid. When Bidder looked, she found that liquid did exist in nautilus.

The observation of cameral liquid was made with the novel use of x-radiography on a living nautilus, although the liquid would have been found by simply breaking open the shell. Radiography was an inspired means of studying the interior of the shell because both the distribution and volumes of liquid in the various chambers could be seen without hurting the animal.

Although Anna Bidder was blessed with a steady supply of nautilus specimens and the first aquarium capable of maintaining

Radiograph of a living nautilus. The soft parts are located in the large body chamber; note the animal's jaws. The dark spots in the lower central part of the body are the kidney stones. Liquid can be seen in the last two chambers.

them, life at Nouméa was not easy for her or for later investigators. The dispute over the origin of the first nautilus radiography is an example. This is how Anna Bidder remembers the discovery of cameral liquid in nautilus (in my judgment the most significant finding ever made about this creature):

> The idea to x-ray a living nautilus came from Dr. Merlet, the head of the medical department in the hospital, who used to come up every evening. He was a close friend of the Catalas and did a lot of the diving for the aquarium. We were trying to think of a way to see if liquid in the chambers actually existed, and Merlet suggested that we try an X ray. Catala got me a nautilus and we put it in a bucket and took it up to the hospital, and I was really rather shocked, because there were all these poor souls waiting for X rays that mattered and

we just went right in and photographed the thing. They developed the photographs, and in one of the older chambers one of the technicians [noticed something and] said, "Isn't that a pneumonia shadow?" And then somebody said, "Let's turn it upside down," and then there was the water in the newest chamber, and the small "pneumonia shadow" of the older chambers had come down. That was the first time we knew there was liquid in the chambers, and that was really historic. I cannot tell you whose idea it was at first, but if Catala is going to claim it all for himself [which he does in his 1964 book *Carnival Under the Sea*], then he is just being very Catala.

There is no doubt that animosity existed between the proud French academic René Catala, the biggest scientific fish in the very small pond of New Caledonia, and the middle-aged zoologist from England. For whatever real or imaginary reason, Gallic pride was wounded, and this rift was to have serious consequences for Denton and Gilpin-Brown of the Plymouth Laboratory, who hoped to go to New Caledonia after Bidder. Of the rift, Anna Bidder made the following observation:

Catala was jealous. He had been very cordial and very welcoming when he met me off the plane. And then, sometime later, in my innocence I mentioned a little paper which said how once Australians had picked up a nautilus from the deck of a fishing boat and kept it in the aquarium in Sydney, Australia, for a week. "Oh," said Catala, "I thought I was the first person to keep nautilus in an aquarium." So I said, "Honey, they didn't keep it alive, they kept a dying nautilus." But it jolted him, and do you know for the whole rest of the time I was there, I was never again invited into their house, until once when he had gone off to Paris, on my very last night in New Caledonia, I said to Stucki [Mme. Irene Catala-Stucki, René Catala's wife], "Come and have supper with me," but she said, "No, tonight you will eat in *our* house."*

*When I stayed at the Nouméa Aquarium in 1975, I heard the Catalas' side of the story. Catching nautilus in New Caledonia is hard, dirty, dangerous work. Most of the time the trapping was done by Mme. Catala and Dr. Merlet, and on the few occasions when Anna Bidder accompanied them out to the reef, she was let off on the lighthouse island while the others voyaged out into the ocean to set or pull traps. They rather unfairly held this against Bidder (who had little experience in boats or the sea) and were sufficiently incensed that they decided never to let English scientists use their aquarium again.

I asked Anna to recount the most important accomplishments of her stay in New Caledonia.

> The really important thing I did was sitting and watching the nautilus in the tanks, and from that I learned a lot about the tentacles, which I summarized in the *Nature* paper. I also did feeding experiments and found out where the food goes in the digestive gland, and very confusing it is. I did some very simple learning experiments and found out that they can learn, and that they learn by touch as well as sight. There is absolutely no doubt at all. When they were newly captured (in baited traps in the sea, then brought to the aquariums in the Nouméa Aquarium) and as soon as they woke up in the evening—for they had a very strong diurnal rhythm [active during nighttime hours and largely motionless during the daytime, remaining attached to the sides of the tanks]—when you went to the tanks in the dusk you would hear *tok*... *tok*... *tok*... [Anna rapped a spoon against her Wedgwood teacup] and they would simply swim along until they would bang into the glass of the tank, and then they would turn until they were alongside and swim parallel to it, until they hit the next wall.
>
> But when they had been in the tanks for quite a few days, they were swimming freely about without hitting the sides. In one of the large square tanks a large gorgonian coral covered one corner, and the nautilus path was just beneath it. One day a large branch of the gorgonian fell and covered the main path of the nautiluses' usual route, and they began to bang into it. I think that this learning must have been partly visual. You see, you can learn a great deal by outline. Living in the blackout of the last war, one learned to use skylines, using the outlines of trees and houses to tell where you were. Such a system would work well with the open pupil of the nautilus eye, and I think they learned the geography of their tank in this way, by silhouette. Some learning in nautilus is clearly from touch, however. As dusk falls and the eyes become less effective, out come the pre- and postocular tentacles, which are normally retracted during the day, and they would feel their way around the tank. In Lifou I made a very beautiful observation. I caught a nautilus and put him in a large portable aquarium. After being put in this tank he extended his tentacles and, feeling the sides of the walls, moved along all four sides of his confinement. Coming back to his original starting place, he settled down on the bottom and retracted his tentacles.
>
> Oh, the waking up in the evening, that was the lovely thing, of course. When they were first in, they would spend a lot of time sleeping on the bottom during the day. Some of my colleagues say I have no right to use the word *sleep*. I can't help it; sleep is what I call it.

And they slept, they slept immobile, with the hood up far enough in the shell to allow the pupil of the eye to be just above the notch in the aperture of the shell, and hence the notch for the eye. Charming, a charming adaptation. And then as the light fell, they would begin to rock a little bit, first one and then another, and for a while I had nearly a dozen in the tank, and then the rocking would get a little stronger, and then off, off, off. And on a clear day, within 20 minutes from all being asleep, they would all be up, like great, lovely, buoyant bubbles. It was a beautiful sight.

But there was also fighting. They would fight over food, and near the end of my stay there were two that were fighting, and I didn't want them to destroy each other because I was running short on specimens, and one had bitten off the entire apertural edge of the other. But on the whole they didn't seem to mind each other and would often sit together, one attached to another, in the corner of the tank. Two or three times I've seen them use their funnels to blow a little hollow out of the sand and sit down on that for their sleep.

Nautilus is different from the coleoids, but not necessarily more stupid. It is slower. It hasn't got to be quick, because it can go into its shell. I think we haven't explored what a nautilus can learn because of this feeling that you must teach by the eye. And I think that the eye is not unuseful, but I am sure that nautilus learns by the touch and taste of things as well.

During her weeks in New Caledonia, Anna Bidder observed every detail of the life and habits of her nautiluses, both in the day and at night. It was probably inevitable that she gave them names, and I am sure that when she found one dead in the tank or had to sacrifice a nautilus to finish a feeding experiment or gather needed tissue samples for study, her remorse was great.

Ever mindful of Willey's quest, Bidder was always on the lookout for the long-cherished fertilized nautilus eggs. This is how she described her observations of reproductive habits:

I got lots of copulation among my nautilus, but it was as likely between two young males as between a male and a female. It was rather funny because I had been doing some social work among the homosexual population over here [in Great Britain], and then I got over there, and one day we separated two copulating nautili and found them to be two young males, and I said, "Oh, no, not that problem here too, I thought I'd left all that behind me!"

Bidder's specific purpose in traveling to the far Pacific was to study the nautilus's digestive system. To do so she collected tissue for later histological study and conducted feeding experiments on live nautiluses. Much of her effort in Nouméa was directed toward better understanding the processes of food detection, acquisition, and digestion.

> The nautilus will eat, of course, anything that smells. They have been caught with dead dog and dead chicken, and crushed crayfish and crushed sea urchin, and dead fish. We fed them with joints of crab and bits of fish, but they would have to have crustacean to get their calcium. All the creatures that make calcium shells have to get their calcium from their food. I think that is why nautiluses are so ravenous on to lobster. It is like a horse going to a salt lick, don't you see?
>
> The food, after it is ingested, can last a very long time in the crop [a large, flexible sac that can expand to hold large quantities of food before it enters the stomach]. I found that the food can sit in this crop for up to five days, and most curiously, after this amount of time is still not decayed. I found big pieces of food that had been bitten and swallowed sitting in the crop after several days and still perfectly sweet and undecayed. But of course the odd thing is, the Japanese have shown that the lining of the crop is not cuticular; the lining of the crop is a mucous membrane. At first I didn't believe them. Then I got someone in the lab to cut me some sections, and sure enough, it is mucous membrane. I wonder if the crop is not secreting some sort of disinfectant to keep the food from rotting. There you have it, sitting in a secreting membrane, and the food does not rot. I don't know, I have no idea. I'm no biochemist.
>
> I did do some feeding experiments, and I thought that I would be clever. I gave one animal three colored meals with indissoluble colors. I gave it a cobalt meal in fish. They didn't very much like fish that had a wedge of coloring stuffed into them. The trick to do was to give the nautilus the colored fish, and then I immediately gave them a clean fish, and that roused all the feeding behavior, and so he swallowed the pill to get at the jam, you see. So I got the cobalt in and waited 22 hours, and then I gave him a green meal, and then I waited another 15 hours, and then I gave him a yellow meal, and then I waited another two hours, and then I killed him. I found all three colors in the ducts of the digestive gland. In another animal I gave a black Indian ink meal and found ink in the absorbing cells of the digestive gland after only 50 minutes. So there isn't any rule; they can store food or use it immediately.

Nautilus with tentacles in food search position. When the animal scents food, its tentacles extend outward in this fashion.

> The thing that fascinated me most about my nautilus was the differentiation of the tentacles into different groups that would come out for different reasons, for different stimuli. Also it seemed that they could learn, and this learning was associated with the use of the tentacles. At dusk you first have the pre- and postocular tentacles [small tentacles on the front side of each eye]. And you also have what I call the alert tentacles, numbers 5, 6, 7, and 8. [A single nautilus has 90 tentacles, which are identified by number.] And they come out like a dog sniffing the air, they come out for information, I am quite sure, they come out for "Hey, what's cooking?" And when food comes near, the next line of later tentacles come out in a lovely sabellalike pattern. And the middle lot, in the front of the hood, only come out when the food is touched by others. They are quite clearly under different nervous control. They always show the same sequence, first the alert tentacles, and then the outer laterals, and only then the middle dorsals. And that I found fascinating, because it was so beautiful! I used to get up on the tank and hold up the food and call out, *"Qui veut manger, qui veut manger?"* and they would all come.

At the end of her three-month stay in New Caledonia, during which she also visited Lifou in the Loyalty Islands, the site of Willey's historic first observation of nautilus eggs, Anna Bidder left for New Britain. She flew to Papua New Guinea and, while trying to take a small plane to New Britain, encountered the chaos one still finds when traveling on Air Niugini:

> Those were incredible flights. You would be told to be ready at half past ten, so you were ready at half past ten, and then they said you hadn't need to be ready before half past twelve, so you were ready at half past twelve, and then they said, "We're getting you lunch and will fetch you out after lunch," and you had lunch and then they fetched you out and then you sat out in the sun for two hours in a little chair with nothing to drink but that horrible green lemonade, and I think there was a little black coffee if you knew where to look.

The New Britain Bidder visited in 1960 was much changed from the one Willey had seen 70 years before. In the 1920s the eruptions of The Mother and Daughter volcanoes increased in frequency and violence, some causing substantial loss of human life. The landscape was changed by these eruptions, and the placid surface of Blanche Bay was on occasion gray with a coating of pumice that made it look as if ice had come to this tropical place.

But these natural changes were nothing compared to the manmade ones. In the 1930s the expanding Japanese empire began fortifying the town of Rabaul on the Gazelle Peninsula of New Britain, and Blanche Bay became one of Japan's largest military bases. The calm bay, with its deep water, was a perfect anchorage for the Imperial Japanese Navy. In World War II the Allies bombed Rabaul and stemmed the tide of Japan's empire. Today, the hundreds of wrecks that line Blanche Bay and rest on its bottom provide a haven for the nautilus to lay its eggs.

Anna Bidder was given a boat, the harbormaster's launch, and, along with a man named Leon, began trapping for nautilus. They were not terribly successful. Anna and Leon would set their traps and then retrieve them after an hour or two instead of waiting the one to three days needed for optimum catches. In six weeks they captured only one nautilus. This specimen, however, led to an important discovery.

> When we caught our nautilus we were so excited that I hardly remembered to do any observations on it. We did make a hole in the

shell, and nothing came out, and a little water went in. The fish that came up with the nautilus died immediately, with a ruptured swim bladder sticking out her mouth, but the nautilus was fine. She sicked up what she had been eating, which was fish fry. What killed the nautilus was the very high temperature, which was 37° C by the time we got back to shore.

In this sequence Anna describes a very important experiment, one that illustrates a major difference between fish and nautiloids. Anna and Leon had pulled the nautilus and the fish up quite quickly from a depth of 600 feet. The rapid ascent caused the volume of gas in the fish's swim bladder to increase rapidly as ambient pressure decreased, a perfect illustration of Boyle's Law of gas expansion. Because the fish was unable to equalize quickly enough, its swim bladder exploded and it was killed. The nautilus, on the other hand,

Trap being towed to trapping site, New Britain, 1960.

which also has air-filled spaces (in the shell), suffered no ill effects. (I have brought nautiluses up from nearly 2,000 feet in a short period of time without their experiencing any undue problems.) What this experiment demonstrated, among other things, is that gas pressure in the nautilus shell, whatever the depth, is always very low. The strength of the shell, rather than gas at high pressure (as in the case of fish), enables the nautilus to maintain the volume of its buoyancy-producing organ. Because of this a nautilus can change depth rapidly, whereas a fish often cannot. When Anna Bidder brought the nautilus she captured in New Britain to the surface, she punctured the shell underwater; had gas been at high pressure in the shell, she would have heard a rapid decompression as the shell was pierced. Instead, water moved into the shell, indicating that the pressure within was less than one atmosphere.

When Anna captured a nautilus in New Britain the local islanders were very disturbed. She had broken a taboo:

> The natives in New Britain were very fussy about nautilus. There were strong taboos. A man who was going to fish for nautilus was to have nothing to do with women, certainly the night before and perhaps longer than that, and no women were to be associated with fishing for nautilus. And so when I went out fishing for nautilus in New Britain, the locals said, "Of course they won't get nautilus, there is a woman on board." There was a delightful priest, he had been in the Vienna Boys' Choir, he had all the Viennese charm. He slopped around in grubby shirts and sort of gray flannel baggy pants and parasol sandals; he was darling. And when we caught our nautilus we rang him up, and he was thrilled because, you see, this broke the taboo. He said, "Yes, my dear, I have heard, we have been *praying* for you." Oh, he was a darling. You know that Viennese charm, there is really nothing else like it.

This was the only nautilus Anna Bidder saw in New Britain. After a six-week stay there, she packed her things and left. With her round-the-world ticket she continued on, heading eastward to Fiji, Hawaii, and the United States before going back to England. She returned to accolades from her colleagues and presented her findings at numerous scientific meetings and colloquiums.

On our last afternoon together in Cambridge, sipping tea, Anna briefly summed up her experiences for me:

> People are so interested in nautilus because you have the whole of the fossil record crammed with things that are so much more like nautilus than anything else alive. The danger, however, is to think that it is a living fossil, when in fact it is a very highly evolved and very specialized animal. And all of these reconstructions that try to shove a nautilus head into an Orthoceras shell [a fossil nautiloid of the Paleozoic era] make me laugh.
>
> Of course there is an aesthetic element, because the shell has been a treasure for so very long.

Finally, I asked Anna what moment in her research had given her the greatest satisfaction, and this was her reply:

> The great moment was when the photographs documenting the various uses and behavior of the nautilus tentacles were delivered to me in my Cambridge lab, soon after I returned from my trip. I took them in to show Carl Pantin of my department, who so long before had set me going on my trip and who had backed me on this trip, and must have wondered very much if I would make a do of it, because nobody quite knew, and I went in and I showed him the photos, and he said, "Anna . . . they *are* good!"

Chapter 5

ON THE BUOYANCY OF THE PEARLY NAUTILUS: ERIC DENTON AND JOHN GILPIN-BROWN, 1962

Once in a while a scientific discovery literally changes the direction of a field of science. Thomas Kuhn, in his classic book *Structure of Scientific Revolutions,* named these changes in perception *paradigm shifts.* These shifts have such a profound effect that they are like a river changing course; all work that follows must take the change into account.

In 1966 such a paper changed ideas about nautilus growth and buoyancy that had been accepted for centuries. Entitled "On the Buoyancy of the Pearly Nautilus," it was written by two British biologists, Eric Denton and John Gilpin-Brown. What is less well known is that Denton and Gilpin-Brown's masterpiece was an extension of their earlier work on the growth and buoyancy of the cuttlefish *Sepia.* But cuttlefish don't have nearly the good press and familiarity nautilus does, so the nautilus paper of 1966 was seen as their classic.

One of these authors, Eric Denton, got me started on my research of nautilus biology. In 1974 I attended a lecture Denton gave to the Department of Zoology at the University of Washington in Seattle. Denton presented his findings on the buoyancy systems of the genera *Nautilus, Sepia,* and the third member of the cephalopod triad still using buoyancy chambers, the tiny squid *Spirula.* I was a graduate student then with an interest in ammonites—fossils with shells similar to those of nautilus. I knew of Denton and his work and was grateful to hear him lecture that day. Afterward I had a chance to speak with this rather shy, diffident man. We were joined by Arthur Martin, a professor of physiology at the University of Washington, who was investigating the function of the kidneys in nautilus. When Dr. Martin mentioned that he was going to New Caledonia the following summer, Denton suggested that I go along to see nautiluses in their natural environment.

The next time I saw Eric Denton was in 1986 in Plymouth, England, where I also met his colleague and coauthor, John Gilpin-Brown.* Denton has been the director of the Plymouth Laboratory, England's most famous marine research station, for more than 11 years. The huge windows of his office overlook the sea at a vantage point from which I imagined many an English sailor, before voyaging to the New World, had watched the Atlantic squalls crash into the Plymouth coastline. We talked about nautilus and the sea and of life in Plymouth in the 1950s, when these two began their studies on the cuttlefish *Sepia,* which ultimately took them to the shores of the Loyalty Islands, the same beaches Arthur Willey and Anna Bidder had visited.

I was extraordinarily comfortable talking to these two men. My work, and the work of many others as well, rests on their broad scientific shoulders. They are also nice people who obviously like each other. Indeed, much of the success of Denton and Gilpin-Brown's research into cephalopod buoyancy mechanisms is due in large part to their harmonious collaboration. The old adage that says two heads are better than one is true only if the two heads can continue to speak together after some period of scientific research. Denton and Gilpin-Brown complemented each other in scientific matters; Denton, a trained physicist, brought a quantitative methodology to a field historically dominated by qualitative measure and observation. John Gilpin-Brown, a trained zoologist, had the biological expertise, insight, and technical skills necessary to pose the tests and experiments that would help them make choices among hypotheses. Equally important in their working relationship was the melding of their humanity, the unpredictable give-and-take of ideas and personalities. I asked them if they had gotten along when they eventually traveled to the South Pacific to study nautilus.

*John Gilpin-Brown died in spring 1987.

"Yes," Denton replied, "[the trip] was fine. The real test of being able to work with someone is whether you can survive an oceanographic cruise together, confined on a research vessel." They have managed to survive many such trips together over the past three decades.

The way these two men began their research on cephalopod buoyancy mechanisms, and ultimately nautilus buoyancy and growth, illustrates the haphazard way science can work. Neither Denton nor Gilpin-Brown had thought much about these animals before their research. Instead, the study of nautilus presented itself through a series of chance events. Denton's first marine research was on the swim bladders of fish, and on a cruise to acquire fish specimens for study he began to think about cephalopods as well. Denton described his pathway to buoyancy research this way:

> I had just been appointed to the Plymouth Laboratory, and I thought that I must work on marine animals. Trevor Shaw [another physiologist at Plymouth at the time of Denton's appointment, in the mid-1950s] and I began to work on the swim bladder of fishes, and it struck me that perhaps this topic was the least interesting problem, and that whatever you found out in the sea that was neutrally buoyant, or at least anywhere near it, must use some mechanisms for maintaining neutral buoyancy, although in many animals there was no actual sign of things. So we decided to look, John and Trevor and myself. Trevor was very keen to work on swim bladders, but I was very keen to work on other things than swim bladders. We decided that if we went to sea and found something that was neutrally buoyant, then we had a problem that was worth looking at. And that is a very good thing on a deep-sea cruise, for when you go to sea you have no idea what you are going to catch, so you need some kind of a hold-on problem that will let you work away on something, whatever it turns out. If you had a very definite plan on a deep-sea voyage in the kind of ship we had then, which was very small, with small nets, you would probably never see the animal you were hoping to work on.

And so on a deep-sea cruise aboard the research vessel *Sarsia* in the mid-1950s, Denton and Gilpin-Brown hauled up in their nets a peculiar-looking squid. When they put it in a bucket of water on the ship, the squid floated freely. Clearly, the squid had some mechanism that gave it extra buoyancy.

Whether an object will float in the sea is determined by easily measured relationships between volume and mass. These relationships were first formulated by the ancient Greek philosopher Archimedes. According to Archimedes' Principle, an object will float if its mass is less than the mass of water it displaces. For example, if the volume of an object in the sea displaces ten pounds of seawater and the object weighs only nine pounds, it will float. A more practical way of determining whether an object will float is by applying measures of density. Density is the mass of an object divided by its volume; the density of seawater is slightly less than 1.03 grams per cubic centimeter. Any object in the sea that has a mass lower than that of seawater for the same volume will float.

The squid Denton and Gilpin-Brown caught and observed was able to float because its overall density was less than or equal to the density of seawater. Squids, however, are normally quite muscular creatures, and muscle (as well as most other animal and plant tissue) has a higher density than seawater. Most tissue varies between 1.04 grams/cc (fatty tissue) to 1.07 grams/cc (muscular tissue). How, then, could the squid float? When Denton, Gilpin-Brown, and Trevor Shaw cut open the squid, they discovered large volumes of liquid that contained high concentrations of ammonia ions. Because solutions enriched in ammonia ions have lower density than seawater, the liquid inside the squid reduced the animal's overall density, giving it buoyancy, much the way a balloon filled with a gas that is lighter than air (such as helium) will buoy up the heavy gondola suspended from it. Over two-thirds of the volume of Denton and Gilpin-Brown's squid was composed of this low-density liquid.

These squids, belonging to the family Cranchidae, spend their entire lives in the middle depths of the sea, far from either the surface or the bottom. They are rarely seen by anything but oceanographers and toothed whales; they are the latter's most common food source. Scientists at the Plymouth Laboratory studying the contents of sperm whale stomachs have concluded that there are enormous numbers of these cranchid squids in the sea. With their ammonia-filled liquid spaces, they are able to keep from sinking into the darkness below. Denton and Gilpin-Brown had made an extraordinary discovery. Considering that they live in the world's largest biotope—the midwater regions of our planet's oceans—the cranchid squids may be the most abundant animals of their size on earth.

Once they had attuned themselves both to squids and to buoy-

Eric Denton (above) and John Gilpin-Brown.

ancy problems, it was perhaps inevitable that Denton and Gilpin-Brown would pause by one of the large holding aquariums in the Plymouth lab basement one day in the late 1950s and notice that many of the cuttlefish in the tank were floating at the surface instead of hugging the bottom, as these squidlike creatures normally do. These specimens were trapped on the surface and unable to descend. Clearly, they had undergone a buoyancy change that had reduced their density and made them lighter than seawater. Denton and Gilpin-Brown had been taking strenuous cruises in search of just such a problem when all along, right in their own basement, the cuttlefish *Sepia* was begging for scientific attention.

A cuttlefish is not a fish at all, but one of the most common cephalopods of the Eastern Hemisphere. The cuttlebone, an internal calcareous skeleton entirely surrounded by the animal's flesh, is commonly found washed up on beaches of Europe, Africa, Asia, and Australia. There are no living cuttlefishes along the coasts of either North or South America (they went extinct there millions of years ago) but they are still familiar to most of us as the white oval objects often placed in the cages of parrots or parakeets. One's first impression when handling a cuttlebone is how extremely light it is for its size. When a cuttlebone is closely examined or broken, hundreds of closely spaced calcareous sheets held together by microscopic pillars of calcite are visible. It has long been known that the cuttlebone gives the cuttlefish its buoyancy. The mechanism leading to the formation of the cuttlebone, however, was a complete mystery before Denton and Gilpin-Brown's research.

Nineteenth-century scientists gave little thought to the cuttlefish and its curious internal skeleton. The early physiologist Paul Bert was one of the few zoologists who studied the cuttlebone. In 1867 Bert published an account of the chemistry of gases in a cuttlebone. He thought that the pressure of gas contained within the cuttlebone would vary with the animal's depth in the sea, much the way a fish's swim bladder works. This explanation seemed correct until Denton and Gilpin-Brown observed that the Plymouth cuttlefish had obviously changed density even though they had not changed depth.

Denton and Gilpin-Brown approached this problem in three steps. First, they wanted to know if the qualitative differences between "heavy" and "buoyant" cuttlefish would hold up under accu-

The cuttlefish *Sepia officinalis,* the species Denton and Gilpin-Brown used in their buoyancy experiments. This specimen had just eaten a crab and expressed pleasure with the tiger-striped patterns on its body. Pictured below is the deep-water cuttlefish *Sepia orbignyana*.

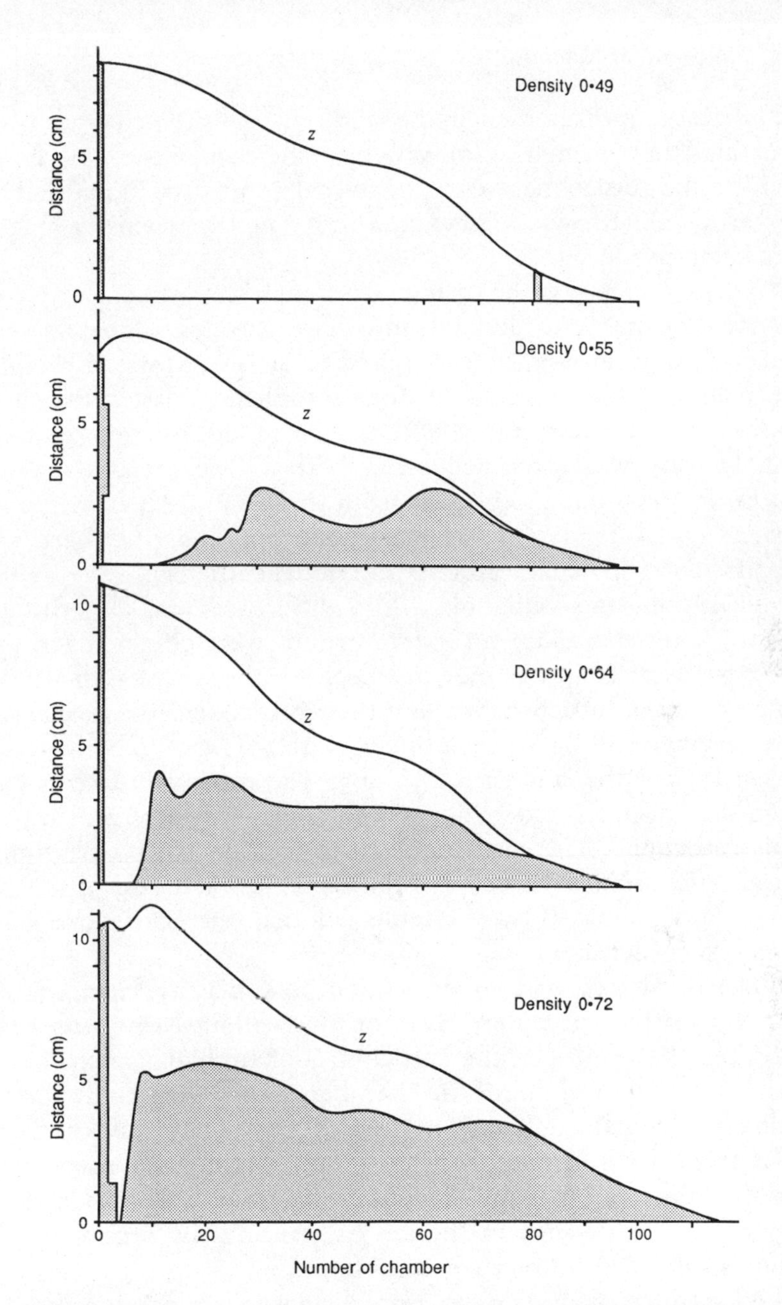

Diagram showing the results that proved that cuttlebone density is controlled by the amount of liquid contained within individual chambers. The ordinate represents distance along the chamber measured from the siphuncular surface, which therefore corresponds to the base line. The inflexion, *z*, marks the thin chambers of very small volume. The regions containing liquid are shown by the hatched areas.

rate density measurements. Second, if density changes were occurring, they wanted to know where the change took place. Was it within the cuttlebone, or in the animal's soft parts? Finally, Denton and Gilpin-Brown wanted to know *how* these density changes were being produced.

The first step was the easiest. Cuttlefish were dredged up from the sea bottom and divided into two groups: "floaters" and "sinkers." Both groups were weighed on sensitive torsion balances, first in air and then in seawater. To accomplish the latter, the cuttlefish were anesthetized and then attached to the balance by a thin wire. Density was computed using both air weight and seawater weight. It was immediately clear from these initial experiments that "floaters" were far less dense than "sinkers," and that there were real differences in density among individual cuttlefish.

Identifying the source of density differences was also straightforward. Cuttlebones from the two groups were obtained and their densities measured just as they had been for the whole animals. The soft parts of the cuttlefish, without their cuttlebones, were also similarly weighed and their densities computed. The cuttlebones varied in density, but the soft parts did not. The density differences between the "floaters" and "sinkers" therefore were determined to be in the cuttlebone. The cuttlebones from the "floaters" were significantly less dense than those from the "sinkers." With this discovery, Denton and Gilpin-Brown established that the cuttlebone alone controlled cuttlefish buoyancy.

The third step was to determine how density changes took place within the cuttlebone. Denton and Gilpin-Brown observed that when they cut open the cuttlebones of fresh specimens, drops of salty liquid flowed out of the chambers. They then knew that the cuttlebones contained liquid as well as gas. They quickly determined that the cuttlebones from the "sinkers" contained more liquid than the "floaters." An age-old puzzle had been solved: Cuttlefish changed their buoyancy in the sea by varying the proportions of liquid to gas space within their cuttlebones.

As so often happens in science, the answer to one question led to another. How could this liquid be removed from a chamber when the cuttlefish was at a depth where the water pressure of the sea would be acting against any mechanism that was attempting to do so? Denton described this stage of the research:

We realized that there was a change in density in the cuttlebone, so we needed to find some mechanism that would allow the pumping of water out of the chambers against the pressure of the sea. At first we thought like everyone else that gas pressure was driving the system; the cuttlefish would drive liquid out of the chambers by increasing the gas pressure within the chamber to a higher pressure than the pressure of the sea [ambient hydrostatic pressure]. And it was a natural thing to think, because we had all been working on the fish swim bladder very recently. And so we did this very straightforward experiment of just puncturing the chambers underwater. We expected bubbles of gas to come out of the holes, but instead the water went in.

This simple experiment entirely changed the understanding of buoyancy in cuttlefish and, ultimately, in nautilus. Denton and Gilpin-Brown had expected that the gas pressure within the cuttlebone chambers would be higher than ambient pressure; when they punctured the chambers, they expected the higher pressure gas to stream out through the hole. Water going *into* the hole indicated that instead of being higher, gas pressure in the cuttlebone was actually *lower* than the one atmosphere ambient pressure at which the experiment was performed.

Denton again: "We had a very perplexing time until we thought of another mechanism—the osmotic mechanism. [Osmosis is the movement of liquids of varying salt concentrations separated by a semipermeable membrane.] And then I produced this mechanism, and Trevor Shaw, who was very, very clever, said that it couldn't possibly be true."

If it was not gas pressure that was pushing water out of a chamber when a cuttlefish became more buoyant, what force was producing this effect? Denton and Gilpin-Brown came up with an ingenious solution to this puzzle. They knew the liquid within the chambers was separated from the Cuttlefish's blood system by a permeable membrane. Consequently, they hypothesized that if cuttlebone liquid had a lower salt content than blood, the liquid would tend to flow into the blood by the process of osmosis. If the salt difference between the chamber liquid and blood was high enough, water would flow out of the chamber and into the bloodstream, even against the force of ambient pressure.

The key to this hypothesis—that the liquid is removed by osmosis—is that two very different forces act against each other. Because cuttlefish tissue is composed mostly of liquid, its pressure is the same as the hydrostatic pressure at the animal's depth. If, for example, the cuttlefish is living at a depth of 33 feet, its tissue is under a pressure of 2 atmospheres. Thus pressure will increase by one atmosphere for each additional 33 feet of depth. At a depth of 33 feet, the cuttlefish must produce some pressure from within the cuttlebone to counteract the hydrostatic pressure and prevent the blood filtrate from flooding the chambers.

Before this discovery, it was assumed that cuttlefish used chamber gas pressure to avoid this flooding. At first Denton and Gilpin-Brown believed that in order for the cuttlefish at 33 feet to keep water out, it simply maintained a gas pressure of 2 atmospheres within its cuttlebone. This theory was shattered forever when they put a hole in a chamber of a living cuttlefish and found very low gas pressure. They determined that osmotic pressure created by salt difference, rather than gas pressure, kept the chamber from being flooded.

To test their hypothesis, Denton, Gilpin-Brown, and the skeptical Trevor Shaw developed an ingenious model to reproduce conditions occurring in cuttlefish chambers. They filled a piece of plastic tubing with fresh water and sealed off its ends with a colloidal membrane that was permeable to water. This membrane mimicked the tissue membrane that separates the liquid in the cuttlebone chamber from the saltier liquid of the bloodstream. They then enclosed this vessel in a larger beaker filled with a salty solution of a chemical called polythene glycol. The salt difference between the fresh water in the small tube and the polythene glycol in the large beaker was so high that it created a very strong osmotic gradient between the two solutions.

Denton described the results of this novel experiment: "It worked like a dream. At the end of the first few hours, a little bubble appeared in the fresh water of the small tube, and then the bubble got bigger, until the small tube was almost empty of water, and then the tiny tube floated up to the surface of the liquid in the beaker." When they examined samples of the bubbles in the small tube, they found them to be almost perfect vacuums. The gas did not force the fresh water out; instead, the fresh water left the smaller vessel through the permeable membrane and in so doing, created a vacuum. As the liquid left, the density of the smaller tube decreased, causing it to float with ever greater buoyancy until the liquid was completely gone. Over a few days' time, gas slowly diffused into the now empty smaller tube.

A new explanation for the buoyancy system of the cuttlefish had been born. It immediately became apparent to the investigators that the osmotic mechanism might be the one used by other creatures as well—including nautilus. For their theory to be accepted as scientific fact, however, Denton and Gilpin-Brown had to demonstrate the correctness of each part of their hypothesis. In methodical fashion, they began investigating the parts of the cuttlebone that were relevant to it.

Their first step was to see if the various chambers acted independently, or in concert. They tested this by puncturing successive chambers of cuttlebones held underwater and measuring the densities before and after the experiment. As the chambers were pierced with a needle, water rushed into each in succession, raising the density of the entire cuttlebone in small increments. Cuttlebones can have more than a hundred individual chambers, and Denton and Gilpin-Brown found that each chamber acts as an independent reservoir of gas and liquid.

The second step was to find out where liquid enters and exits the chambers. The two scientists put a freshly dissected cuttlebone in a vacuum and lowered the pressure. When tiny beads of liquid emerged from the chambers, from the edge where they rest against a sheet of blood-filled tissue called the siphuncle, they determined that each chamber had only one site through which liquid could pass.

The third step was to analyze the chemistry of the chamber liquid. If the mechanism proposed by Denton and Gilpin-Brown was correct, the cuttlebone liquid had to have a lower salinity than blood. They measured the salinity in several different chambers from a number of cuttlefish and compared these values to the salinity of the animals' blood. In every case, the chamber liquid was less salty than blood.

Finally, according to their hypothesis, the cuttlefish changed density by increasing or decreasing the amount of liquid in the chambers. This was already known to be correct: The cuttlefish with higher densities—the "sinkers"—had higher volumes of liquid within their cuttlebones than the "floaters," which in some cases were almost empty of liquid.

Although Denton and Gilpin-Brown's osmotic hypothesis had

passed every test, a number of questions still remained. If gas pressure within the chambers was always lower than atmospheric pressure, why wasn't the cuttlebone crushed by the hydrostatic pressure of the sea? And how did liquid get into the chambers in the first place?

The first question was easily answered. Denton and Gilpin-Brown observed that cuttlebones had a number of distinct adaptations that increased their strength, such as the presence of pillars between the chambers. Such adaptations kept the cuttlebone from being crushed, at least in shallow water.

The question of chamber liquid was more difficult. Denton and Gilpin-Brown had to explain both the observed volumes of liquid found in various chambers, and the observed salinities of these liquids. Their answer was ingenious. As the cuttlefish grows, naturally it must increase the size of its cuttlebone; to accomplish this, it must create a new chamber on top of the last one formed. Because the liquid that initially fills each new chamber comes from the bloodstream, it has the same salinity as blood. When the new chamber space, still completely filled with this liquid, is sealed off by the calcification of a new chamber wall, the buoyancy system becomes functional. The end of the new chamber has a slitlike opening that is in contact with cuttlebone tissue. This tissue begins to actively remove ions of sodium and chloride from the chamber liquid, gradually lowering the salinity. When the liquid's salinity drops to the point where its osmotic difference from blood produces a pressure that equals, or just surpasses, the force of hydrostatic pressure, liquid will leave the chamber and move back into the bloodstream. For cuttlefish in a surface aquarium, the liquid will begin to exit the chamber when its salinity is about 0.9 that of the blood (which itself has a salinity of seawater). At ever greater depths, progressively more salt must be removed before emptying will take place. Denton and Gilpin-Brown derived an equation that gives the relationship between salinity and depth; a cuttlefish living at about 200 feet, for example, would have to lower the salinity of its chamber liquid to about 0.7 the salinity of seawater for emptying to take place.

The step-by-step elucidation of this beautiful adaptation was slowly worked out in the tanks and laboratories of the Plymouth marine station, with each step in the unraveling puzzle patiently demonstrated by the classical scientific method of hypothesis testing. Each tenet of the entire theory was demonstrated by repeatable experiment.

In 1959, after Denton and Gilpin-Brown had patiently worked out the physical principles of the cuttlefish's buoyancy control system, they began a series of experiments to see how the animal actually used this system. In a remarkable study, they demonstrated that cuttlefish change their density in response to changing light intensities. Up to this time the collaborative work of these two investigators seemed to have been dominated by the physicist's classical methods, but the experiments demonstrating the effect of light on the buoyancy system were the inspiration of the zoologist part of the team.

Denton and Gilpin-Brown were originally drawn to their study of cuttlefish because of the distinctly different buoyancies these animals exhibit. In their natural habitat, cuttlefish are nocturnal. They have evolved the habit of burying themselves in the sand during the day to protect themselves from predators; at night they leave their shallow burrows and hunt for crabs. Denton and Gilpin-Brown wondered if this cycle could be associated with buoyancy change. In their 1961 paper on this subject, they wrote: "The behavior of well-fed cuttlefish is greatly affected by light. When the light is bright they usually bury themselves in the gravel at the bottom of their tanks, whilst after twilight they come out of the gravel and swim around until dawn. Here we show that striking changes in buoyancy accompany changes in light intensity."

Denton and Gilpin-Brown documented the effect of light on cuttlefish buoyancy by periodically weighing the animals during alternating light and dark cycles. Weighed in seawater during periods of light, the cuttlefish were found to be denser and less buoyant, and they buried themselves in the gravel. When the lights over the tanks were turned off, the cuttlefish became less dense and more buoyant, and they emerged from the gravel to swim around. In fact, specimens kept in the dark for many days continued to increase in buoyancy over the entire experimental period.

In a series of 1961 papers Denton and Gilpin-Brown demonstrated that the presence or absence of light, rather than some other environmental stimulus or internal regulating factor, was responsible for the buoyancy changes they observed in cuttlefish. They were also aware that this system might have relevance to other cephalopods with chambered shell portions. "The cuttlebone," they wrote in their initial paper on cuttlefish buoyancy,

> has a special interest in that it is closely related not only to the shells of the living *Nautilus* and *Spirula,* but also to the shells of the fossil Nautiloidea, Ammonoidea, and Belemnoidea. These animals dominated the Paleozoic and Mesozoic seas and their evolution from a crawling to a free-swimming life was probably determined by the use of the shell as a buoyancy device. We hope in the light of new knowledge of the cuttlebone to see more clearly how these important animals lived. The chambers of the cuttlebone contain liquid as well as gas, and the cuttlefish changes its density and posture by varying the amounts of liquid which the chambers of the bone contain. If the fossil cephalopods could also have done this, their behavior must have been very different from that postulated on the usual assumption that their chambered shells were completely filled with gas.

Denton and Gilpin-Brown were fairly sure that the cuttlefish's buoyancy system was also used in some way by nautilus. The irony was that the cuttlebone, which is contained entirely within the animal, is a highly modified design of the flotation device originally evolved by the Cambrian nautiloids. The question then became: To what extent do the two systems differ? Only with a voyage to the Pacific in search of nautilus would they be able to find an answer.

Denton and Gilpin-Brown had finished their experiments on the buoyancy system of the cuttlefish by the time Anna Bidder set off for New Caledonia and New Britain to observe nautilus. Before departing, she visited the Plymouth Laboratory to confer with Denton and Gilpin-Brown. She was primed in advance of her trip about the possibility of chamber liquid in nautilus, and as we have seen, she did discover it. But her findings about the buoyancy of nautilus were not conclusive. And so the two experimentalists from Plymouth had to see for themselves.

By 1961 Eric Denton had lost his collaborator to a university post in New Zealand, but he decided to travel to New Caledonia to make the experiments about nautilus buoyancy that would either confirm or refute the applicability of their cuttlefish findings to it, and planned to conduct his field work with Gilpin-Brown. The

Royal Society awarded Denton a fellowship of £1,000—hardly enough to cover expenses for such a trip.

Arriving in New Caledonia, Denton encountered his first serious problem: Dr. Catala, the director of the Nouméa Aquarium, refused him permission to carry out the nautilus experiments there. The success of the cuttlefish studies had depended largely on being able to keep the experimental animals alive for various lengths of time in aquariums. Without such facilities, it would be difficult to conduct the hoped-for experiments on nautilus properly. Thus impeded and limited in resources, Denton was strapped for solutions. He described the problem as follows: "We went in with the idea that with any luck the nautilus buoyancy system would behave like that of the cuttlefish. If it didn't we would be absolutely sunk. We had very little time, and very little in the way of equipment, and if nautilus hadn't behaved like the cuttlefish we would have gotten nowhere." After this first brief visit to New Caledonia, Denton traveled to New Zealand to meet up with Gilpin-Brown.

Denton and Gilpin-Brown arrived in New Caledonia together in November 1962.* Because of the difficulties with the Catalas, they were uncertain how they would procure nautilus there, so they decided to follow Willey's lead and conduct their experiments in nearby Lifou. Unfortunately, Lifou had no scientific laboratory of any type, and Denton and Gilpin-Brown would have to capture their animals and make the various measurements either in tide pools or in the large plastic garbage pails they had brought along with them. Denton was understandably concerned about the possibility of not catching any nautiluses:

> My main worry was that I had taken £1,000, which seemed like a lot of money to me at the time, and [might] go around the world and come home with bugger nothing in my hand. I was in conflict because I had never seen a nautilus. So we designed an experiment which was guaranteed if it worked to give something publishable from one animal. It was a combination of experiments in which we did some weighings and some puncturing of the shell and some histology, and extracted some gas. It wasn't perfect in any respect, but perhaps if we

*In our interview, Denton and Gilpin-Brown's reminiscences of this trip were aided by the diary Gilpin-Brown kept while they were in the South Pacific that year.

> got one animal, and we were lucky and nautilus really did operate like *Sepia,* then you had enough to know something useful. John had already done some experiments on *Spirula* [a deep-water squid also using the osmotic system of buoyancy] and out of the goodness of his heart was prepared to share his results with me if we got no nautilus.

The two scientists arrived in Lifou in mid-November. In every nautilus searcher's account of his journey to Lifou, there is one familiar observation: It is a beautiful island. Lifou has a series of upraised limestone reef cliffs, stark white rock against a riot of green vegetation. The climate is tropical, and the island is surrounded by coral reefs even more splendid than Australia's Great Barrier Reef. Denton and Gilpin-Brown stayed in a three-room house, already the home of chickens and pigs, with the latter, by Gilpin-Brown's account, living in the bathroom. According to Gilpin-Brown's journal, after a first meal of pigeon and spaghetti, the two scientists went to bed amid "scratchings and scrapings of a big green bug, scuttling land crab under the head of the bed also, one cockroach, and the crickets outside."

Denton and Gilpin-Brown, true to their European promptness, began their nautilus trapping operations immediately. They spent most of their first afternoon, however, waiting in vain for a local Lifouan who had promised to assist them with their first trapping. When he failed to arrive, they discovered that life in the Pacific was governed by a sense of priorities quite different from their own: Why do something today if you can do it just as well tomorrow or, better yet, next week? With the afternoon almost gone and no help in sight, Denton and Gilpin-Brown decided to try trapping by themselves.

Neither man was a novice at sea: Both were veterans of oceanographic cruises, and were familiar with boats large and small from their work at Plymouth. Their Lifouan research vessel, however, must have been a shock. John Gilpin-Brown described their craft and first attempt at setting traps for nautilus:

> The boat proved to have no floorboards, it is clinker built, and has no motor, but unpaired oars, one of which was nearly worn through. Since I had been in the navy for three years, Eric immediately expected that I should row. It was a beast of a boat. The breeze made quite a slop on the sea, which made handling the boat difficult [and

> this is probably a serious understatement on Gilpin-Brown's part: a 20-knot tradewind usually blows in the afternoons in Lifou, making the sea very rough]. Two traps placed in rather shallow water, and came back at dusk.

Denton and Gilpin-Brown spent the first of many nights wondering what their catch would be. Like nautilus workers before and since, their thoughts that night gradually moved from how the experiments would turn out to possibly not catching a nautilus, which would mean having no experiments to worry about at all. The next morning Denton and Gilpin-Brown rowed to their traps, pulled them up from the depths, and, with hearts undoubtedly pounding at the prospect of finally seeing a living nautilus, found nothing. The traps were empty. So they reset them and rowed back to shore. As the heat of the day increased, the two investigators had an afternoon to pass.

This was their daily ritual: row out to where the traps were set and again in the tropical heat make the backbreaking effort of pulling them up empty from 100 feet below. The two investigators, so primed to see finally if nautilus did indeed act like *Sepia,* had to spend these days in make-work projects, passing the hours until they again went out to sea. They began collecting nautilus shells and cuttlebones found washed up on the beach, perhaps thinking of what could have been, or might yet be. Gilpin-Brown's diary also mentions whiskey before supper.

After many days of unsuccessful trapping they decided to try deeper water, a difficult prospect because the bottom in Lifou drops off to very great depths over a short horizontal distance, and they had no fathometer. Finally estimating that they were over a good depth, Denton and Gilpin-Brown put the wood- and chicken-wire trap into the water and payed out the line. The trap sank into the cobalt-blue sea. At the end of the rope they threw the buoy into the sea, and then watched in horror as the buoy sank, never to be seen again. The depth had been too great for their length of line. John Gilpin-Brown did not record the mood of the returning sailors in his journal. On the following day, armed with new traps and a longer line, they placed their devices in the sea, waited the night, and the next morning pulled up their hopes. The bait within the traps had been eaten, but still there was no nautilus.

On the morning of November 22, Denton and Gilpin-Brown

were awakened by the sound of torrential rain pounding on the corrugated iron roof of their house. The two investigators decided they had better get their traps. John Gilpin-Brown's diary: "Awoke to sound of pouring rain. Breakfast of two cups of coffee and sallied forth. Very overcast and windy, had a hard row against the wind, managed to retrieve only one trap, but no nautilus. Overtaken by unpleasant squall, very heavy rain, and fresh wind. Thoroughly soaked. Left remaining traps and returned for shelter. Returned to house at 9:30 A.M., rubbed down, and drank rum." The two biologists tried to recover their other traps that afternoon, but were thwarted by the howling wind and sea.

The next morning Denton and Gilpin-Brown awoke to gentle breezes, a calm sea, and a sense that the nightmare of the previous day was finally over. Their boat, however, was full of water and had to be bailed before they could search for their traps. They had no problems finding the traps and had them on board by 7:30 A.M. —but for the fifth day in a row, without nautilus. Although the diary gives no clue as to their feelings, the two investigators must have been somber after such hard work and no reward.

On their sixth day of deep-water trapping, Fortune finally smiled on Denton and Gilpin-Brown. With a solitary nautilus in their possession, they returned to shore in triumph, only to find several Lifou islanders waiting for them. On the previous day, with no nautilus, and no assurance of ever catching one, Eric Denton had promised to pay the Lifouans 1,000 francs (about $10) for a living nautilus. As Denton recounts, a Lifouan named Noel was the first to greet them. "'You did say you would give me 1,000 francs for a nautilus,' said Noel. I said, 'Yes,' and he said, 'Suppose I caught two, would you give me 2,000 francs?' and I said, 'Yes,' and he said, 'Suppose I caught three,' and I said, 'Yes, 3,000 francs," and he said, 'I caught seven!'" In one fell swoop, eight nautiluses were in the large plastic garbage can, not necessarily happy, but alive. Seven thousand francs changed hands, and Denton and Gilpin-Brown set to work.

Noel and his companions were able to capture the nautiluses because of a break in the weather. Although the latitude of New Caledonia and the Loyalty Islands is such that trade winds blow most of the year, the months of November through March are the typhoon season, when raging hurricanes slide southward from their equatorial spawning grounds. In between these tropical depressions there can occur extraordinary periods without a hint of wind, when the sea becomes a flat mirror. Such a calm coincided with Denton and Gilpin-Brown's stay and allowed the Melanesians to practice a unique form of nautilus fishing. On these moonless nights the islanders rowed out beyond the reef and let down fishing lines with large hunks of fish on the ends. They then waited for a telltale pull on the line and ever so gently raised it, luring the nautilus to the surface, where, still holding the bait with its tentacles and nibbling, the animal was scooped up in a waiting net.

To ensure that some of the nautiluses would stay alive, Denton and Gilpin-Brown put the animals in closed-off traps that were then placed in large tidal pools at the edge of the sea, serving as aquariums. Other traps with nautilus in them were put back in the sea at various depths. Having hedged their bets against unforeseen accidents, the tired but happy scientists, according to John Gilpin-Brown's diary, "returned home rather late after dark to find the island's only doctor already installed. We had a large dinner, including soup, vegetables, fish trimmings (fins and eyes, etc.), fish, steak, pigeon stew, salad, more vegetables, and dessert."

With animals in the bank, so to speak, Denton and Gilpin-Brown no longer had to go to sea and could begin their experiments immediately. They were finally free of the pressure of finding nautilus; at last, as Denton told me, "I felt relaxed." They did row out daily to check and feed their captives in the various holding traps, but work now shifted to their makeshift laboratory, where the first experiments began to confirm their expectations on the nautilus buoyancy system. Meanwhile, Noel, their now-enthusiastic Melanesian assistant, brought in seven more nautiluses two days after the first group. They paid him another 7,000 francs, and the word was out: The two mad Europeans were paying exorbitant prices for live nautiluses. Denton remarked that he began to hope the new trade in nautilus on Lifou would diminish a bit, because he was running out of money. John Gilpin-Brown's diary records that on the night of November 27, "supper was rather overshadowed by arguments about the price of nautilus."

Denton and Gilpin-Brown received only one more nautilus, captured by their Lifouan friends shortly before they left. This nautilus, however, was more valuable than all of the rest combined because it was their only immature specimen. By sheer luck this specimen had recently produced a new chamber, which was clear from the paper-thinness of the last-formed septum. Here was a chance to solve the question of how nautilus produces a new chamber: Was the new chamber space filled with gas, as Lankester and Willey had believed, or filled with liquid, as suggested by their own studies of cuttlefish in Plymouth? Gilpin-Brown's diary records the answer, based on the first actual observation: "In afternoon we saw our first specimen with a relatively new chamber. Partition [the calcareous septum that closes off a chamber] thin and chamber full of liquid." In this one short sentence the problem was resolved. New chambers are initially filled with liquid, not gas. The preseptal gas hypothesis of Willey and Lankester, cited for an entire century in hundreds of invertebrate zoology texts, was as dead as the idea of an earth-centered solar system.

Although Denton and Gilpin-Brown conducted their experiments in Lifou with limited equipment and no aquarium facilities, they were able to design a number of tests aimed at specific questions about nautilus buoyancy. They had brought with them sensitive torsion balances like those used for the cuttlefish to weigh living nautiluses in air and underwater and compute the densities of various specimens. They also had brought chemical fixatives that enabled them to preserve soft tissues. Their decomposition thus retarded, the tissues were embedded in plastic, sectioned with a diamond knife, and later observed and photographed with an electron microscope. Denton and Gilpin-Brown were also able to seal up the nautilus shells, preserving the liquid and gas within the chambers for later analysis.

With living nautiluses now available, Denton and Gilpin-Brown began their work on buoyancy. They set up experiments and observations that would answer the following questions: Is nautilus neutrally buoyant, and therefore weightless in seawater, or is it either lighter or heavier than seawater? What is the mechanism of buoyancy change? How is a new chamber formed?

Their first step was to measure the buoyancy of a freshly captured nautilus. The completely grown specimens of *Nautilus macromphalus* weighed about two pounds in air. In water, however, they weighed about 1/25th of an ounce. The weight of the nautiluses in seawater was about 1 percent of their weight in air—almost, but not quite, perfect neutral buoyancy. Denton and

Gilpin-Brown anesthetized some specimens and found that they slowly settled to the bottom of their holding tank. Weighing the freshly captured nautiluses in seawater revealed an important fact: Although they were nearly weightless in the sea, they still had a slight weight. None of the specimens weighed less than seawater; in other words, they were not positively buoyant, which meant they would tend to float to the surface. This important principle is well known to scuba divers. If a diver has insufficient weight on his weight belt, he tends to float to the surface. It requires a lot of energy for the diver to kick hard enough to compensate for this positive buoyancy and drive himself downward. On the other hand, weight sufficient to cause a slight negative buoyancy that allows the diver to sink is much less fatiguing. A diver's fins are made so that a slight kicking motion pushes him forward and slightly upward. A diver with a bit too much weight expends much less energy staying at a prescribed depth than a diver with too little weight. Nautilus has the same problems. A nautilus swims by forcing water out of a fleshy tube that extends out of the front of the shell. This tube is located *beneath* the mass of the tentacles, and can never point vertically upward. Because of this, the tube can more easily drive the nautilus upward than it can push the nautilus downward.

Denton and Gilpin-Brown also discovered that as a nautilus grows, it must enlarge both its shell and its body. While the flesh enlarges at a nearly constant rate, the shell enlarges through the production of calcium carbonate in two different areas. The shell's overall size increases as the nautilus secretes new calcium carbonate along the apertural lip, which thus enlarges. As the lip gets bigger, more room is created in the body chamber, the portion of the shell that houses the animal's soft parts. A new chamber is formed when the body chamber enlarges to a point where the animal has room to pull away from the last-formed wall. This process, unlike the production of new shell at the apertural edge, occurs periodically rather than constantly.

How, then, does the nautilus maintain its neutral buoyancy while going about these complex operations of shell formation? Denton and Gilpin-Brown had an ingenious answer for this question. The scientists weighed the soft parts of several nautiluses in both air and seawater and computed their densities. They found slight, but significant, differences in these. They then measured the density of each animal's shell. This figure varied in such a way that they found the combined density of shell and animal to be approximately equal to that of seawater. They discovered that the nautiluses with the highest-density soft parts had shells of the lowest density.

Their next question was, What causes differences in shell density? A nautilus might be able to regulate the amount of calcium carbonate produced in its shell, but this process would not allow for rapid density change. Denton and Gilpin-Brown found their answer within the chambers themselves: Nautilus shells of high density contained more liquid in their chambers than the lower-density shells. To confirm this, the two scientists emptied all liquid from the chambers and again measured shell density. Without liquid, all the shells had the same density.

In 1966 Denton and Gilpin-Brown announced their discovery: "The differences between the densities of the living tissues were compensated in the whole animal by the shell's containing differing amounts of water. It thus seems fairly certain that nautilus can regulate its density by varying the amount of water within the shell and that it does so to make its density slightly greater than that of seawater."

Since the time of Robert Hooke in the 1600s, scientists had thought that the secret of nautilus's buoyancy, and buoyancy change, depended on the amount of gas in its shell. Denton and Gilpin-Brown showed that chamber *liquid,* not gas, was the key element.

New questions arose from this discovery. How did liquid get into the chambers, and once it was there, how was it removed? The cuttlefish findings provided an important clue. When a cuttlefish produces a new chamber, liquid at first fills the newly created space and is later removed osmotically. Denton and Gilpin-Brown traveled to Lifou convinced that the nautilus used a similar system. Their suspicions were confirmed when they discovered that the juvenile nautilus's newly produced chamber was filled with liquid rather than gas. They also found that this chamber liquid had the chemical composition of a bodily secretion and was not seawater. The next step for Denton and Gilpin-Brown was to show how the liquid is removed.

The chamber of a nautilus shell is very different from that of a cuttlefish. The nautilus's chambers are independent of one another, but they have a thin calcareous tube running through them. This tube, the siphuncle, provides the motive force for buoyancy change in nautilus. It is composed of two distinctly different parts, an outer, calcareous part and an inner, fleshy one. The outer part of the siphuncle is made up of a material resembling chalk and, like chalk, is permeable to water. The inner part of the siphuncle is composed of tissue with many blood vessels. Anna Bidder had earlier observed a sample of this tissue under an electron microscope, and had found that it closely resembled the tissue in the cuttlefish that is responsible for pumping water out of the cuttlebone. Prior to their voyage Denton and Gilpin-Brown had suspected that the siphuncle was probably the source of buoyancy change in nautilus. The simple tests thcy performed in Lifou showed that the siphuncle was the only place where water could enter or leave a completed nautilus chamber.

If the siphuncle was the "pump" that removed liquid from a chamber, what was the mechanism of that pump? Once again, the cuttlefish results provided a starting point. In cuttlefish, water leaves a chamber when a salt imbalance exists between liquid inside the chamber and the blood within the animal's siphuncular tissue. Nautilus has a similar, though highly modified, system. The blood vessels within its siphuncle carry an adequate supply of blood at seawater salinity. The chamber liquid is in contact with the blood through the porous siphuncular wall. If the chamber liquid could be shown to have a lower salinity than the blood, the osmotic model would work for nautilus as it did for the cuttlefish.

When Denton and Gilpin-Brown removed liquid from the chambers and discovered that it indeed had a lower salinity than blood, they knew that their hunch had been correct and that osmosis was the answer to buoyancy change in nautilus. In only one specimen was the chamber liquid's salinity close to that of blood, and that sample had come from the immature animal in the process of new chamber formation.

Denton and Gilpin-Brown were finally able to summarize the processes of new chamber formation and buoyancy change in nautilus: "These observations strongly favor the hypothesis that a new chamber is formed by the secretion of some body fluid between the animal and the inner wall [last-formed septum] of the living chamber. This body of liquid is then sealed off by a new septum and a new length of siphuncular tube, and it is only when the new septum and the siphuncular tube are sufficiently strong to withstand the pressure that the liquid within the chamber is pumped out."

Denton and Gilpin-Brown, who conducted their nautilus experiments largely in wastebins and tide pools, completely changed the understanding of the creature's buoyancy system. One great mystery still remained, however: What was the role of the gas within the chambers?

For centuries it had been thought that variation in gas pressure was responsible for nautilus buoyancy change. Even after making the discovery of chamber liquid in *Sepia,* Denton and Gilpin-Brown thought at first that variation in the pressure of the gas that was also present was what the cuttlefish used to force liquid out of its chambers. This hypothesis had to be discarded, however, when they found that the gas pressure was always low. Their studies of nautilus confirmed that chamber gas was not used in buoyancy regulation. Denton and Gilpin-Brown found that the pressure of gas from within nautilus chambers was always less than one atmosphere. They also found that the younger the chamber, the lower its gas pressure. For instance, removing liquid from a newly forming chamber creates a virtual vacuum. By sampling a series of chambers from the same animal, Denton and Gilpin-Brown found that it takes many months for gas to rise to its maximum pressure of slightly less than one atmosphere. Gas enters the chambers as it comes out of solution in the blood. It is not actively produced, or pumped into the chambers at varying pressures. Gas, therefore, plays no part in buoyancy. If anything, the introduction of gas into the space vacated by the liquid *reduces* buoyancy (though imperceptibly), since gas has weight. The strength of the nautilus shell, not gas pressure from within it, guards against implosion from hydrostatic pressure.

Measuring gas pressures from single shells gave Denton and Gilpin-Brown their final and perhaps most ingenious finding. Because the gas from successively older chambers was found to increase in a regular fashion, the two scientists were able to make a rough computation of the relative age of each chamber. They produced a curve plotting gas pressure against time and came to the following conclusions: "It can be seen that the points on the curve corresponding to the chambers are distributed as if the chambers were laid down at a roughly constant rate, and that on our present hypothesis the time interval between the formation of successive chambers is one about [every] 13 days."

Since an adult *Nautilus macromphalus* usually has about 30 chambers, this estimate suggested to Denton and Gilpin-Brown that full growth would be reached in about a year. It was still not known how long a nautilus lived after that. Here was the first scientific calculation for growth rates of nautilus, based on the brilliant insight that characterizes the work of these two scientists. Unfortunately, this estimate and the premise of constant periodicity in the rate of chamber formation would later prove to be very much in error.

Denton and Gilpin-Brown were in Lifou for less than a month. They recall spending their last days observing the beautiful coral reefs surrounding the island. They feted their Lifouan friends at a farewell supper, followed by another feast in Nouméa, before heading back to New Zealand, their notebooks crammed full of the numbers and observations that would forever change our understanding of nautilus. In New Zealand they performed the crucial chemical analyses on both the chamber liquid and gas. Denton then returned to England. Like Willey, neither of the two ever set out in search of nautilus again. And like Willey's, their findings sowed the seeds that sent more voyagers out into the waters of the Pacific to study this elusive creature.

Chapter 6

THE LIFE SPAN OF A NAUTILUS: ARTHUR MARTIN AND THE NOUMÉA AQUARIUM, 1970–1975

In his poem "The Chambered Nautilus," Oliver Wendell Holmes offered the first estimate of the nautilus's life span. He based the poem on his opinion that each year a new chamber is added to the shell:

> *Year after year beheld the silent toil*
> *That spread its lustrous coil;*
> *Still, as the spiral grew,*
> *He left the past year's dwelling for the new,*
> *Stole with soft step its shining archway through,*
> *Built up its idle door,*
> *Stretched in his last-found home, and knew the old no more.*

With their delightful sense of style, Denton and Gilpin-Brown used the passage above to close their classic 1966 work on nautilus growth and buoyancy, remarking that Holmes's estimate was "out by an order of magnitude." The two scientists had arrived at an estimate of about 14 days between successive chamber formation events, which they based on their measurements of gas pressures from within successive chambers. According to Denton and Gilpin-Brown's estimate of two weeks per chamber, the nautilus would reach full growth about 300 days after hatching. However, they were slightly uneasy about this estimate, as shown in this passage: "This growth rate, which gives a doubling in weight about every 40 days, seems very great, but cephalopods do grow quickly; thus, Denton and Wilson have found similar values for *Sepia officinalis* (the English cuttlefish) in its second summer, while in its first summer the growth rate is even faster. The period of about one year for complete growth is moreover not an unlikely one for a cephalopod."

The one-year growth rate Denton and Gilpin-Brown proposed was not derived from actual observation of shell or tissue. Without aquarium facilities they were unable to make the long-term studies of actual shell growth that would have helped corroborate the estimate they based on the function of gas pressure. Had Denton and Gilpin-Brown been able to observe the physical development of an immature nautilus in the Nouméa Aquarium, they would soon have realized that their proposed growth rate was far too high.

During the two decades following Denton and Gilpin-Brown's estimate work, a new cadre of explorers trekked to the Pacific to measure nautilus growth rates. Three questions concerned them.

First was the actual time needed for a nautilus to reach sexual maturity and growth cessation after hatching. This was of vital interest to paleontologists concerned with the growth rates of the extinct nautiloids and ammonoids as well as to physiologists, although for a very different reason. Denton and Gilpin-Brown had demonstrated that in nautilus and *Sepia* a new chamber is originally filled with liquid and that this liquid is almost completely pumped out before another new chamber is produced. Growth in nautilus was therefore linked with the osmotic mechanism of chamber liquid removal, a system that piqued the curiosity of cell physiologists in the late 1960s and early 1970s.

The second question about growth was how long a nautilus lives *after* it reaches maturity. Almost all cephalopods grow very quickly, reproduce, and then die. Nautiluses, however, are so unlike most other modern-day cephalopods that no scientist could be sure this pattern held for it.

Finally, those studying nautilus wanted to know the rate of growth during different stages of development, or ontogeny. Denton and Gilpin-Brown assumed that the formation of each new chamber occurred between approximately equal time intervals. Were they correct?

THE BIRTH OF THE NOUMÉA AQUARIUM

To study growth rates (and much else about nautilus) it was vital to have a supply of freshly captured specimens and a way of keeping them in cooled seawater aquariums. As Willey, Bidder, Denton, and Gilpin-Brown knew, the best place for this study, then as now, was the New Caledonia region because of its cooler water temperatures. And it was on the island of New Caledonia that the Nouméa Aquarium came to exist, in 1956. It was just bad luck for the English investigators that this superb marine facility, perfect in every way for studying nautilus, was established and run by an Anglophobic Frenchman.

The Nouméa Aquarium is perched on one of the most spectacular pieces of real estate in the world. One can sit on the grounds and watch the sun set into the Pacific behind the distant barrier reef 12 miles offshore. The Aquarium is a short walk from miles of sandy beaches. Even today, with many new modern aquariums all over the world, nothing compares with Nouméa's. It is neither large nor lavish. It boasts no swimming pool–sized aquariums or circular shark tanks or dancing dolphins. Instead, its great claim to fame are the long-lived corals that grow within its tanks, shared with numerous fish and invertebrates.

In the 1930s René Catala, a naturalist and farmer on a large Madagascar plantation, was approaching middle age. Although his original scientific interest was the study of butterflies, his focus completely changed in 1934 after a visit to France. While stranded on his way home in a small coastal Madagascar town, Catala had his first view of the underwater world, and it changed his life and plans.

"Dividing my time between the forests and the sea," he recalled,

> I decided to rent the services of some Malgache fishermen in dugout canoes to go fishing. The first fish that I pulled out were *Thallosoma,* superb fish of the warm seas. I was so embarrassed to kill these beautiful jewels that I put them back into the sea, to the extreme disapproval of my native fishermen, who shouted, "But we eat them!" One of these men dived into the sea soon after this to free the anchor and chain from the coral below, and wore a pair of goggles composed of

> simple pieces of glass placed in flexible wood. It was with these rustic little goggles that I first discovered the fairy world of the coral gardens. These brief incursions into this totally new world gave me a whole new orientation to my scientific curiosity.

Catala decided to open a marine laboratory for the study of the coral reef world off the Madagascar coast. Dreams, however, are sometimes shattered by unexpected events, and as so often happened in the days before Fansidar and chloroquine, René Catala contracted malaria and began to suffer recurrent debilitating attacks. Catala did eventually establish a small aquarium in Madagascar, but the increasing frequency and virulence of his malarial attacks ended his project, and nearly his life.

During one of his frequent hospital stays in the late 1930s, Catala was given this advice by his doctor: "You have malaria in its most serious stage. If you wish to regain your health in a country both healthful and exotic at the same time, and realize this dream that you carry in your heart for a grand study of marine organisms, I suggest you go to New Caledonia, where I was a colonial doctor in 1905. Because of the great richness of marine life there, I guarantee that you will be completely satisfied." In his feverish state, Catala fixed on the idea and never looked back. It was at this moment that the Nouméa Aquarium was born, though it would be almost 20 years before the dream became reality.

In 1937 Catala went to France to convalesce. During this three-year period he completed his Doctor of Science degree, based on his observations of butterflies in Madagascar. He arrived in Paris during the onslaught of World War II, when any attempt to raise money for a marine base in faraway New Caledonia was certainly out of the question. While Catala was sitting out the war and continuing his planning, he met an extraordinary woman of an extraordinary time: Irene Stucki, a beautiful Swiss from an old, established family who, in addition to money, possessed a will of iron and the ability to outwork most men. During the war she worked for the French Resistance, and seemed at first a strange match for the sickly colonial naturalist. But a strong bond existed between them—a love of the sea and of the creatures within it.

As the war neared an end Catala began knocking on every scientific and governmental door to enlist aid for his project. In 1945, however, the reinstalled French government had far greater things to worry about. When Catala finally received some official encouragement from the government (but no money), he married Irene Stucki in late 1945, and they immediately set out for New Caledonia.

The Catalas stopped in America, where they visited with many scientists and marine biologists who made them feel welcome and supported. After a long Pacific crossing, the Catalas arrived in New Caledonia just as the last American soldiers based there were heading home. The Americans left behind uncounted tons of matériel, buildings, and property, all of it there for the taking.

René Catala was a man of tremendous vision and had far more in mind than simply a marine aquarium. He wanted to build a scientific establishment where research could be conducted on both marine and terrestrial aspects of the tropical environment. His main obstacle was a lack of money. But happily, the means and facilities were scattered everywhere around New Caledonia.

Catala soon acquired enormous quantities of surplus American equipment: glass, electrical cable, tools. He also immediately made friends with the remaining American authorities, because he had his eye on an even larger prize: a huge suite of buildings located on one of the island's largest beaches. This encampment had served as a hospital for the American forces and included living quarters for staff as well. After long negotiation with the America authorities, Catala received the 70 buildings and tons of equipment associated with "Hospital 105." Catala's long-lived dream seemed about to be realized.

His newly founded Institut Français d'Océanie was composed of four parts: a botanical section for the study of indigenous plants and agriculture in New Caledonia; a farm school to teach New Caledonians new methods of agriculture; a mining institute; and a marine laboratory. Now that the Institut was in place, France began sending scientists to man it. France also decided to take an active role in planning and deciding the direction of further research. Incredibly, it then decided that the new institute, located within a few meters of the Coral Sea, did not need the aquarium facilities Catala required for maintaining the marine animals he so desired to study. France thanked Catala for his superhuman efforts in establishing the Institut, and then ignored him. He was no better off than he had been back in 1937.

With great bitterness Catala and his wife returned to France in

René Catala, 1949.

1952. They had decided they would build an aquarium in New Caledonia at any cost, and toured various European aquariums to gain ideas and expertise for their eventual design. But they were shocked by the backwardness of the so-called "modern" aquariums. None of the facilities had open seawater systems; all used closed systems in which the same water was recirculated over and over. In every case, fish were the principal attractions, and a sad lot they were. Catala resolved that his facility would take large volumes of water directly from the sea, circulate it through the exhibition tanks, then return it to the sea. Instead of the poor electric lighting he encountered elsewhere, Catala decided to use only natural sunlight to illuminate his tanks. And most radical of all, he decided that the centerpieces of his aquariums would be invertebrates. If he could maintain a healthy invertebrate population, he reasoned, he could be assured that the accompanying fish also would be healthy.

In 1953 the Catalas returned to New Caledonia and combined their family fortunes to buy land on the sea and build their aquarium. Construction of the open seawater system was the largest part of the project. The Catalas extended long pipes 300 feet offshore to a depth of 20 feet in the nearby Baie des Citrons. These pipes created a system that brought tons of water into the aquarium each day that was then passed back into the sea. The tanks were without lights and open to the sky above. In 1956 the Catalas opened their aquarium to the public without great fanfare. An entirely new concept, it had neither curio shops nor live animal shows, but was instead consecrated to the exhibition of invertebrates as well as to fish. The aquarium's inhabitants were provided with superb living conditions that, for the first time, all but eliminated the need to constantly restock animals. The habitants of the Nouméa Aquarium became the Catalas' extended family, a large family that would grow and prosper.

With his aquarium finally in place, René Catala and his wife hired a team of divers to gather as many invertebrates as possible from the vast lagoon and barrier reef surrounding New Caledonia. They then began to observe and describe the habits of the many species captured and placed in the aquarium. Much of the Catalas' early work involved finding the best means of feeding and caring for the vast array of invertebrates and fish that filled the tanks.

The stony corals were most difficult of all. In most aquariums, scleractinian and alcyonarian corals must be replaced constantly because the polyps that make up the colonies usually feed on microscopic animals and dissolved organic material not found in closed systems. Even Nouméa's open system was insufficient. The Catalas solved this problem by giving their corals a "milkshake" each morning—a protein-rich soup they made by macerating clam soft parts in a blender, then placed into each tank. The result was spectacular coral growth. Never before had coral grown naturally in an artificial environment.

Concerned as they were with invertebrates, the Catalas of course wanted a nautilus for their aquarium. Their cue may have come from Anna Bidder's initial exploratory letters. They received their first nautilus from local fishermen. Catala wrote in his book *Offrandes de la mer:* "It was the 15th of May, 1958, that we received our first nautilus. A date especially memorable for us, because we didn't imagine, at that time, the great interest the scientific world would take on finding that the conditions offered by our aquarium for conserving invertebrates could also maintain nautilus in good health for long periods of time. At this time we also could not predict the remarkable work on this animal that would ultimately be undertaken in our aquarium. For us, this first nautilus was *the* grand curiosity."

Catala had opened the aquarium for his own study of marine organisms, but he also allowed other investigators to conduct experiments in a well-equipped back room. Who was invited to the aquarium was clearly up to Catala and his wife: Nouméa was not a public facility; it was a privately financed institution. Thus he was able to pick and choose his guests as he saw fit.

In 1960 Catala permitted Anna Bidder to study nautilus in his aquarium. The two did not get along, to the detriment of Denton and Gilpin-Brown, who were subsequently refused the opportunity to work there. In fact, until Arthur Martin arrived in 1970, no one had conducted nautilus research at Nouméa since Anna Bidder.

ARTHUR MARTIN AT THE NOUMÉA AQUARIUM: 1970, 1972, 1975.

In the late 1950s and the 1960s Arthur Martin, a professor of zoology at the University of Washington in Seattle, conducted a series of observations and experiments on the function and structure of cephalopod kidneys, using the giant octopus common to the cold, estuarine waters of nearby Puget Sound as his chief subject. Martin's work led him to wonder, like many other cephalopod specialists before him, how the nautilus would compare. Martin had corresponded with Anna Bidder about her trip and knew about Denton and Gilpin-Brown's problems in obtaining permission to conduct their research at the Nouméa Aquarium. He was therefore pleasantly surprised when Catala answered his request to visit with a hearty yes.

Catala still felt great affection for Americans, whose generosity had helped launch his first scientific endeavors in New Caledonia. Perhaps he also thought it was time for further nautilus research in his aquarium. Following Anna Bidder's visit, Catala had published a hugely successful book, *Carnival Under the Sea* (1964), and after that made a film with the same title. Illustrated with Catala's magnificent photography, both richly described the life and habits of Nouméa's inhabitants, and nautilus was the star.

When Martin first visited New Caledonia in 1970, the island was undergoing a remarkable economic boom. Around that time incredible quantities of the island's rich nickel and cobalt deposits were being mined and smelted, causing a huge "nickel boom" that led to runaway inflation and created difficult living conditions for the thousands of French and Pacific peoples streaming into New Caledonia. Suddenly millionaires were commonplace. Gas-station owners who sold fuel to the trucking companies began driving Mercedes-Benzes; local bakers lined up to buy yachts. In less than ten years, the local Melanesians were no longer the majority in their own country. Today they are still the minority.

Science boomed as well. France increased its scientific presence in New Caledonia, and Catala's original Institut Français d'Océanie became a center of the Office de la Recherche Scientifique et Technique Outre-Mer (ORSTOM), that country's overseas oceanographic establishment. When Arthur Martin arrived, he descended on a city crammed with people searching for lodging and paying exorbitant prices for services and goods.

Martin had come to New Caledonia to research a complex physiological problem: the role of the kidneys in the nautilus's metabolism, especially in its growth. Martin knew that nautilus kidney systems had structures and elaborations not found in other cephalopods, and he longed to know why.

Martin was particularly puzzled by the thousands of tiny granules of calcium phosphate in nautilus's kidneys, whose presence had long been known but never studied. These kidney stones, shaped like tiny concentric balls or rods, have a composition somewhat similar to the calcium phosphate of human teeth. Sometimes as much as two grams of these kidney stones are present in one nautilus.

Why would nautilus have such large quantities of kidney stones when almost no other cephalopods had them? Are they dietary in origin, the result of too many meals of crustacean? Are they simply storage for unwanted mineral salts? Or do they have a more subtle function? Martin had an important clue: The cuttlefish, which also has a complex calcareous shell with buoyancy chambers, is the only other cephalopod with kidney stones. Martin reasoned that there might be a connection between the kidney stones and some aspect of the buoyancy systems in these two animals. He hypothesized that kidney stones were a reservoir of calcium that could be carried through the bloodstream to the site of shell building during periods of new chamber formation. He traveled to Nouméa to test this idea.

By the time Martin arrived, the Nouméa Aquarium had been open for 14 years. Large corals had spread over the sides of many tanks, and fish brought in during the first days of operation in the mid-1950s had become gigantic. Martin was in his mid-fifties then and was just completing his long tenure as chairman of the Department of Zoology at the University of Washington, which had grown into one of the most respected in the world under his leadership. Martin was taking a well-deserved rest and returning to active research after years of administrative and budgetary battling. But his year's stay at the Nouméa was to be anything but restful.

By 1970 the administration and day-to-day operation of the aquarium was handled by Catala's wife, who used the name Irene Catala-Stucki. Approximately Martin's age, she was a striking woman, tall and strong, and devoted to her husband and the aquarium. Much of her typical day was spent cleaning tanks, feeding animals, and running the bookkeeping, as well as maintaining her home and entertaining the constant stream of politicians, scientists, and French celebrities who visited this most exotic New Caledonian tourist stop. She also did most of the scuba diving for the aquarium, collecting specimens both inside and outside the barrier reef. Catala by this time was happily spending his days taking photographs of his beloved animals and writing his second book. In his late sixties and once again in failing health, he made no great efforts to talk with Martin unless absolutely necessary, but relied on his wife to accommodate the visiting scientist in his research. She spoke perfect English, after all, and had one of the sharpest tongues imaginable.

The techniques for capturing nautilus had come a long way since Willey's time. Three-foot cubical traps of wood and chicken wire were baited with fish or crab, attached to surface buoys, and put into the sea at depths of about 300 feet for one or two nights. Soon after his arrival Martin was building such traps, a laborious project in the hot tropical sun. He was then introduced to a local fisherman who, with an ancient wooden fishing boat equipped with

Arthur Martin making buoyancy measurement, Nouméa Aquarium, 1975.

a 5-horsepower diesel motor, took him to the trapping areas on the outside of the barrier reef that surrounds New Caledonia 12 miles offshore. It took three to four hours to traverse the large lagoon enclosed by the reef; the round trip, plus the time spent setting the traps and hauling them up, had to be accomplished in the morning hours to avoid the trade winds that sprang up each afternoon. Martin would meet the fisherman at 3:30 A.M. and on good days he would be back in his laboratory by noon.

Nautiluses were extremely plentiful during Martin's visit in 1970. Recovering a dozen specimens from each trap was not uncommon, and the animals did beautifully in the aquarium's tanks. Martin began his painstaking operations on the captive nautiluses, catheterizing the various parts of the kidney during delicate microsurgery. He began to unravel the function of each component and preserved a great deal of material for further study.

In one respect, however, he was stymied. His physiological studies of the role of the kidney stones were not giving him understandable results. Martin very much wanted to know if during chamber formation some chemical was produced that dissolved the kidney stones so that their calcium could be used to form shell. Unfortunately, he had no way of knowing when new chamber formation was taking place. According to Denton and Gilpin-Brown's growth-rate estimate, this should occur every two weeks. The actual time needed for a new septum to form, therefore, would have to be much shorter than two weeks. Martin thought the calcification would take place in a matter of days. To better understand this process, he began to measure nautilus growth rates so as to determine the periods between new chamber formation.

Martin found an ally in this new research. Stucki, as she liked to be called, had always been in her illustrious husband's shadow, but here was a project she could contribute to, a project that appealed to her sense of wonder about the animals of her adopted home. She began to take an interest in Martin's periodic weighings and shell measurements. Soon she was returning to the aquarium each night, when the strongly nocturnal nautili fed. She made sure each specimen ate.

Catala appears to have been pleased by his wife's newfound interest:

> As soon as the nautiluses arrived in the aquarium, my wife took charge of them, looking after their feeding and health. How many times did I hear, coming back to our house each night, "They are all in good health, but three are being difficult, and two are making me tear my hair out," and for these several individuals, which one would today call "difficult characters," my wife would return each night to

Trap with captive nautiluses inside, New Caledonia, 1975. This trap design was used by the Catalas and Arthur Martin.

The holding aquarium Arthur Martin used, 1970–1975.

> the aquarium often at very late hours to be sure that these naughty subjects would decide to eat, and accept some food sooner than some other. In other words, these nautiluses must not be characterized as dumb cattle, as they must be administered to according to their character and individual taste.

Martin and Stucki began to study the growth of captured nautiluses by measuring the amount of new shell produced at the shell edge each week, and by weighing the nautiluses in air. They started on the smallest nautilus they could find. Unfortunately, for unknown reasons, capturing very small nautiluses was impossible. The smallest specimens available were already in the last third of growth, with between 20 to 25 septa out of an eventual total of 30.

Both the shell measurements and the weighings were subject to unavoidable errors. The new shell material a nautilus secretes is at first paper-thin, and this fragile leading edge is often broken in the aquarium, either by other nautiluses or by contact with the vertical sides of the tanks. If any part of the apertural edge is broken, it must be repaired before any further growth can continue. The specimens Martin and Stucki observed had numerous scars on their apertural edges, indicating that such breakages were very common.

Martin and Stucki found that a nautilus's weight varied according to how much food it had eaten during the week prior to being put on the scale. Because nautiluses are scavengers, they are never sure when they will see their next meal and will usually eat as much as possible of what is offered. A growing nautilus can eat as much as 50 grams of fish at a time, which would clearly show up in the weighing. If nautilus had not eaten for a week or more, it would show a net weight loss. The weight measurements were therefore useful only if taken over many months.

Martin and Stucki kept their nautiluses in a large outdoor tank filled with fresh seawater that was pumped out of the nearby lagoon. Although perfect for keeping corals alive, this seawater system was the downfall of the nautilus experiments. The collaborators began their growth observations in October 1970; by December they had seen significant shell growth and weight increases in the five nautiluses they used as subjects. In January, however, the nautiluses began to refuse food, and in early February they began to die. The reason was the increasing water temperature. The study had begun in the Southern Hemisphere's spring, when the temperature of the seawater pumped into the aquarium was 24° C. As spring gave way to tropical summer, the water temperature rose; when it reached 27° C, the nautiluses sickened and died. Although Martin and Stucki had collected three months' worth of data on growth, summer water temperatures stymied further experiments.

Martin's next trip to New Caledonia was in 1972. This time he began his growth experiments in the austral winter months of July and August. Seven immature specimens of *Nautilus macromphalus* were captured, weighed, and measured. These specimens grew until December, when they too began to sicken, and finally died early in 1973. Martin did manage to get about five months' worth of growth data this time, but he still was not satisfied with his results. He decided to make one final trip to New Caledonia in July 1975, and for the first time, he took an assistant to help him in the great labor of nautilus capture. This was the trip Eric Denton suggested I take.

Early on the morning of August 6, 1975, I flew out of Fiji, which is an hour and a half from New Caledonia by air. As the plane rose above the Fijian coast I had my first glimpse of the huge reefs so characteristic of the tropical Pacific. Coming into New Caledonia, I saw an even larger reef, in size second only to the Great Barrier Reef of Australia. I was astonished by the distance of the New Caledonian reef from the island itself and by the giant lagoon it creates. Even though I had spent the last 18 hours in flight or between flights, when my plane rolled to a halt at Tontouta Airport, I was tremendously excited.

Within an hour of landing I was deposited at the doors of the Nouméa Aquarium. Arthur Martin had preceded me by a week and was nowhere to be found. I wandered around the outside of the aquarium and finally found a back entrance. Dragging the bags and diving gear I had brought for my intended three-month stay, I was met by a large, silver-haired woman with a beautiful smile. She introduced herself in French as Mme. Catala-Stucki. She asked if I was tired, which in my foolish befuddlement I denied. Happy to find an American so refreshed after a day's flight across the Pacific, she put me to work shucking clams for fish food. It was several hours later when Martin rescued me.

Martin and I lodged in a small bungalow near the aquarium, across the street from one of the most beautiful beaches in the world. Early each morning we would walk to the aquarium, gather our bait and traps in a Land Rover, and drive to an ancient stone quay to meet our ancient fisherman. His boat had a small diesel motor and sails, was *very* slow, and looked as old as he did. We would tie the traps on and begin our long voyage out to the barrier reef in complete predawn blackness. My arrival on New Caledonia coincided with the rainy season, and I soon discovered that as long as I could ignore the stench of diesel and dead fish, I could just fit under a tarp covering the engine and get in another two hours of precious sleep.

I had always dreamed of the South Pacific as a paradise of calm, warm water under a tropical sun. I soon learned that the unceasing trade winds blowing from the south created a sickening short chop within the lagoon. We had to travel 12 miles across this mad lake to reach the barrier reef, and then slowly make our way through one of the reef's passes to the outer slope, where nautilus lives. Going through these passes was always an exhilarating experience. Because the tidal change causes great volumes of water to move through the passes four times a day, these areas are sites of great biological productivity and are filled with fish. We usually entered the pass at sunrise, just in time to watch the feeding schools of tuna and mackerel, with their air escort of screaming birds, or see the fins of tiger and hammerhead sharks as these creatures left the lagoon after a night's feeding. Dolphins and whales were common; our boat always chased schools of flying fish into the air, and occasionally we would startle manta rays into making great leaps from the sea.

As we passed into the blue expanse of the Coral Sea, we would begin searching for our red buoys. All too often they would be gone, taking with them our hopes and expensive gear. The traps were massive affairs of chicken wire and wood, weighted with blocks of cement and attached by yellow line to the surface buoys. If the winds came up at night, the waves would pull on the buoys, often causing the traps to dislodge from their resting places and float away. On one occasion Martin and I set five traps outside the reef, only to have a huge storm blow out of the south that night. When we finally got back to our traps two weeks later, the trapping area was bereft of any trace of our endeavors. We found one lone buoy some miles away, crocheted to the reef by a strand of broken rope.

Because the traps had to be weighted as much as possible to avoid drifting, pulling them up was backbreaking labor. Martin and I took turns hauling on the thin polypropylene line from the bow of the boat. We had no winch or depth finder. After 15 or 20 minutes a white blur would appear in the clear water, signaling the arrival of the trap from 50 to 100 meters below. We would scramble to look down into the water, hoping to catch a first glimpse of moving white circles within the trap that would represent a successful catch.

At the beginning of our trip, Arthur Martin was a tall, sturdy man in his early sixties who still had a great thatch of blond hair. During the first month our catches of nautilus were low because of continual foul weather. Much of our time was spent constructing new traps and acquiring specimens. After a great rainstorm at one point, the water being pumped into the aquarium turned fresh, killing the few nautiluses we had. Soon after this episode Martin's hair quickly began to turn gray. I became increasingly concerned that the strain of our failures was getting to him. Over a meal accompanied by marvelously cheap but delicious Bordeaux, I tried to console him and asked him about his hair. He laughed in his wonderful way and told me not to worry: His supply of Grecian Formula had just run out.

By September we had enough nautiluses to start the growth experiments again. As in previous years, the specimens were periodically weighed and apertural shell growth measured. A major drawback continued to be our inability to know when a new chamber was forming in the nautilus. Although it would have been possible to determine this by x-raying our specimens at the local hospital as Anna Bidder and Catala had done more than a decade before, doing this on a regular basis was impractical.

When we added the results of these 1975 observations—conducted from July through December—to the results from previous years, we discovered that the growth rates of the *Nautilus macromphalus* specimens kept and observed in the Nouméa Aquarium were far slower than Denton and Gilpin-Brown's 1966 two-week-per-chamber estimate. Martin, Catala-Stucki, and I presented three estimates in a paper we published together in 1978. Although we could not observe new chamber formation, we did see possible 30-day cycles in weight fluctuation, which we thought might coincide with the formation of a new chamber space. Since most *N. macromphalus* have 30 chambers, this estimate would indicate a growth period of about 30 months, assuming a constant interval of time for each chamber.

Estimating growth rates by weight changes was indirect, however, and not altogether reliable because weighing values were strongly influenced by the dietary habits. The second estimate we proposed was derived from measurements of daily apertural growth. In the seven specimens we observed, a mean rate of 0.25

mm of new aperture was secreted per day. Assuming this rate remained constant through ontogeny, full growth would be reached within five to six years. We corroborated this estimate by counting shell growth lines. From our observations we knew that one shell growth line, a thin ridge of shell material paralleling the aperture, was secreted about every other day. By counting all of the growth lines on a shell, we found that the time from hatching to maturity might take as long as six years. Assuming that the chambers were also produced at an approximately constant rate throughout ontogeny, these estimates indicated that a new one was formed every six to eight weeks, not the two weeks Denton and Gilpin-Brown had proposed.

It was quite a shock that the first actual measurements of nautilus growth gave rates far slower than expected. Our belief in Denton and Gilpin-Brown's estimate was so strong that we decided there must be a reason why growth was so much slower in the aquarium than in nature. We considered the effects of shell breakage, dietary differences, high temperature, or difference in water chemistry between an aquarium environment and the deep-water habitat of the nautilus. Data are never wrong; only their interpretation by flawed human scientists can be in error. We were correct in assuming that nautilus growth rates observed in the aquarium were different from those of animals in their natural habitat. What we didn't know was that growth rates in the aquarium were *faster,* rather than slower, than the natural ones.

THE CHANGING OF THE GUARD

The 1975 season was the last one Arthur Martin spent in New Caledonia. Like Willey, Art Martin was stymied in his original quest: He never found the chemical reaction that would tie the kidney stones to new chamber formation and calcification. But also like Willey, he was resourceful enough to move to other studies when his original questions had to remain unanswered. His work on growth and buoyancy change were fundamental contributions to the understanding of nautilus.

As for Catala, he also left the aquarium soon after our 1975 visit. Increasingly frail and sickly, Catala relinquished the aquarium, after protracted and acrimonious negotiation on his part, to the city of Nouméa. It ultimately came under control of Catala's old nemesis, France's ORSTOM, the original scientific complex he had established soon after the war. After more than 20 uninterrupted years the Nouméa Aquarium changed hands. The new director, Yves Magnier, immediately closed the facilities for changes and modernization. When it reopened some weeks later, the aquarium was young again, with new paint and tanks, and a curio shop—anathema to the Catalas—to help defray expenses. When Catala closed the door for the last time, he never returned to his aquarium, even though he still lived a few short steps away. Perhaps in empathy, the giant Napoleon fish, dean of the aquarium and a 20-year resident, rolled over and died. Nouméa grieved briefly and then, as life does, it went on.

Chapter 7
GROWTH RATE DEMONSTRATED

MOON HISTORY IN A SEASHELL was the headline in *Science News.* "The earth-moon distance deduced from a mollusk!" declared *The New York Times*. Science writers scrambled to get the story. On the cover of the October 19, 1978, issue of *Nature,* one of the world's most prestigious scientific journals, was a photo of a nautilus shell. In the long lead story two young scientists named Peter Kahn and Stephen Pompea claimed a major new astronomical discovery, presenting evidence that suggested the moon's orbit was enlarging at a rate 17 times faster than previously published figures had indicated. The moon, in effect, was fleeing from the grip of the earth's gravity far more rapidly than anyone before had believed. Their evidence for this claim came from a most unexpected source—nautilus.

At the time of Kahn and Pompea's discovery, the American public was hungry for readily comprehensible science. In the mid- and late 1970s this demand was met by several new popular scientific magazines; *Discover, Science 78,* and *Omni* all began to compete with the long-established and venerable *Scientific American* for the lay reader's dollar. They all needed arresting scientific stories, and the more sensational the better. Here was a tailor-made subject: the recession of the moon from the earth as deduced from a seashell.

Kahn and Pompea's thesis was simple. They had observed an average of 30 growth lines on a nautilus shell in the space between two septa. They assumed that these growth lines were produced daily, indicating that a new septum was produced each month.

Kahn and Pompea then reasoned that new chamber formation was triggered by some aspect of the lunar cycle. To support their idea, they used the article Martin, Catala-Stucki, and I had then recently published reporting on our New Caledonian observations.

Kahn and Pompea had visited various museum collections and counted the number of growth rings between septa on fossil nautiloids from different geological ages. Their startling conclusion was that the older the fossil nautiloids were in geological age, the fewer the number of growth lines they had between successive chambers. Kahn and Pompea found, for instance, that nautiloids of the Devonian age, which occurred some 400 million years ago, had about 20 lines per chamber. If the growth lines represented days, it might be deduced that the lunar month during the Devonian period was 20 days long. This could have been true *only* if the moon had been much closer to the earth than it is now, a fact no one has ever seriously disputed. What remained a matter of dispute was exactly *how* close the moon had been at a given time in the past.

Peter Kahn, a graduate student in paleontology at Princeton University at the time, was first struck with the idea while he was examining collections of fossil nautiloids. He saw—or thought he saw—a constant number of growth lines between successive chambers in fossil nautiloids from several different geological ages. Kahn knew that other scientists, notably Keith Runcorn, a British physicist, and John Wells, an American paleontologist, had successfully shown through their studies of the growth bands on fossil corals that there was a greater number of days per year in older geological periods. Kahn's use of nautiloid growth-line data was merely an extension of this type of work. But his intuitive jump to thinking about the earth-moon distance was the hallmark of brilliant science. With the help of Stephen Pompea, a graduate student in astronomy, he worked out a final model.

In 1977, Kahn and Pompea asked Arthur Martin and me for advice about their model before their article was to be published in *Nature*. We pointed out the flaws we saw in their reasoning. First, our best nautilus growth data (the actual measurements of apertural shell growth) suggested that *longer* than a month elapsed between chamber formation events. Second, even at that time Martin and I were pretty sure that chambers were *not* produced at constant intervals, but that each new one took longer to form than the last. Kahn and Pompea listened to our objections and countered with points of their own. They felt that their geological data curve, produced by measuring the number of growth lines between chambers in successively older nautiloids, was too regular to be the result of chance. Despite our objections, they decided to publish their findings.

Reaction in the scientific community to the Kahn/Pompea article was immediate and clamorous. *Nature* printed a number of letters objecting to the hypothesis on a variety of grounds, and then, apparently feeling rather burned by the whole subject, defensively told its readers that no more space was available for additional objections (which kept pouring in anyway). Two groups of scientists carried especially sharp knives: the astronomers, who couldn't believe that a paleontologist could possibly correct their astronomically derived estimates for the earth-moon distance through time, and zoologists, who were interested in growth lines in other invertebrates. Those closest to the story, the paleontologists and others interested in nautilus itself, were lost in the shuffle.

The Kahn/Pompea argument rested on four assumptions that had to be true for both ancient and modern nautiloids. First, growth lines had to be secreted daily. Second, the 30 growth lines had to be secreted between two successive events in chamber formation. Third, new chambers had to be formed at constant intervals. Finally, all fossil nautiloids from the same time period had to show the same number of growth lines between chambers. Kahn and Pompea were eventually proven wrong on all these assumptions.

Both of these young scientists (neither yet had his Ph.D. at the time their article appeared in *Nature*) were stunned by the overwhelming negative reaction. Kahn withdrew from Princeton and moved to Berlin to continue his studies. Pompea began teaching at a small college in Colorado. Overlooked in the controversy was the fact that two scientists from vastly different branches of science had joined forces to produce a truly cross-disciplinary piece of research. This melding of sciences foreshadowed a similar alliance between two disciplines five years later when astronomers and paleontologists would successfully team up to study extinctions at the Cretaceous-Tertiary boundary.

The greatest benefit of Kahn and Pompea's brilliantly conceived hypothesis was that it focused attention on the problem of growth rates in nautilus. New research was the result, particularly in these areas: What is the relationship between the growth of new shell at the aperture and the formation of new chambers? Are new chambers produced at a constant rate, or at increasing or decreasing rates as nautilus ages? And, most important of all, what are the real growth rates of nautilus in nature?

THE PROCESS OF NEW CHAMBER FORMATION: NEW CALEDONIA, 1979

In 1979 I returned to New Caledonia to continue work on nautilus. Since my first visit there with Arthur Martin in 1975, I had spent three months in Fiji (in 1976) and several weeks in both Fiji and New Caledonia (in 1977). All of these trips had been financed by small grants from my Ph.D. supervisor, Dr. Gerd Westermann of McMaster University in Hamilton, Ontario, or from my employer at the time, the Ohio State University. The 1979 expedition was financed by a grant the National Science Foundation awarded jointly to me and to Lewis Greenwald, a brilliant physiologist at Ohio State. Although our primary research mission was to deal with questions of buoyancy, we had questions about growth very much on our minds.

The beauty of a federally sponsored grant is that you have adequate equipment and financial support with which to conduct your research. My previous trips had been on very tight budgets: airfare, money for food and lodging, but little or no money for scientific equipment, and next to none for what was most important of all—boat time. In the past I had resorted to haunting the grocery stores and bars nearest the yacht harbor, hoping that conversations with newly arrived skippers would lead to invitations for a sail. If I wasn't immediately thrown out when I showed up with my ropes and nautilus traps, I was assured of another supply of fresh specimens.

In 1979 Lewis Greenwald and I leased a 20-foot powerboat and equipped it with a winch and a depth sounder. We also had two new tools that revolutionized our ability to study nautilus: a portable X-ray machine, and tanks in which we could store chilled seawater that would enable us to keep our nautiluses at about 15° to 20° C, the same temperatures they encountered in their natural habitat. The tanks, large fiberglass affairs painted black to keep the light out, had the size and appearance of two large coffins. Dr. Magnier, the new director of the Nouméa Aquarium, had installed them specifically to better maintain nautilus specimens, and the creatures thrived in them.

Our boat, which we had christened the *Arthur Willey,* turned out to be a deathtrap. Although supposedly reconditioned at one of New Caledonia's marinas, the *Willey* tried (and often succeeded) in every way imaginable to strand us at sea. On several occasions we motored all the way from the outer reef to Nouméa with only our 5-horsepower "security" engine. On one such mission this too failed. The *Willey* also had the nasty habit of refusing to start if its engine had been running for some time and had then been turned off (or quit on its own). To start it again we would have to wait until the Volvo Penta engine cooled, then remove the air filters from the two carburetors, squirt ether into the intakes, and pray it would start before the battery died.

On one occasion Greenwald and I had spent the day motoring between distant trapping sites, and it was almost dark when we began to set our last trap just offshore of the crashing reef breakers. As usual the wind was blowing, and we had great difficulty getting the heavy trap over the side of the boat in the swell. The engine was running all this time because we didn't dare turn it off for fear it would never start again. While both of us were putting the trap over the side, the engine decided to die. Lewis and I both heard the telltale sputtering. I dived head first onto the front seat and jammed the throttle wide open. The engine sighed and then, a breathless moment later, caught. Without an engine we would have been washed onto the reef in moments. Though not necessarily fatal, spending the night on an exposed reef in heavy surf is not an experience I would want to go through. We were both very shaken during the two-hour run back to Nouméa. This trapping adventure, however, yielded one of our best catches and let us spend more time conducting science and less time on boat work.

We had sufficient nautiluses in our cooled seawater system to begin long-term growth observations. We x-rayed our specimens once a week and soon realized that the chambers were being produced at a rate much slower than once a month. The nautiluses also grew much more slowly at 16° C than they had in the warmer temperatures of the 1975 experiments; growth rates in nature were clearly far *slower* than those in the aquarium.

As the weeks stretched into months, our periodic radiography of immature nautiluses began to reveal a fascinating tale. The X-rays showed not only the interior of the shell, but also the water levels within individual chambers. Suddenly we could watch the many simultaneous growth processes in a nautilus. We found that the formation of new shell material at the aperture was continuous, and that even if it was halted by small breaks or by poor feeding or water quality, the nautilus still usually secreted a new increment each day. After weeks of watching this process, we concluded that the average rate at which new shell material was produced was about 0.15 mm per day, much less than the 0.25 mm of new shell material produced by the nautiluses we had kept in the warm-water tanks in the early 1970s.

The radiograph also enabled us to observe new chamber formation. Until then, one of the major questions about nautilus growth was how the back of the body pulled away from the last-formed septum during new chamber formation. Was the septum secreted quickly, and then did the body gradually retreat afterward, thereby creating the liquid-filled space that would become the next chamber? Or did the back of the body rest against the last-formed septum for a long time, then move forward suddenly in a period of days or even hours? The radiographs from our patiently growing nautilus proved that the latter was the case.

Even more fascinating was the beautiful interplay of apertural shell growth, chamber liquid removal, and the timing of new chamber formation. Our radiographs showed that a nautilus began the initial step of creating a new chamber—moving its body away from the last-formed septum and leaving a liquid-filled space behind it—at precisely the point when the last-formed chamber was half emptied of liquid. The half-empty mark in the chamber was easily visible in the radiographs, and over the weeks we were able to watch the liquid slowly disappear. The position a living nautilus assumes is such that the convex swell of the septum in the last-formed chamber is oriented like an umbrella. The siphuncle, the internal tube that empties the liquid from the chamber, sits vertically atop the hemispherical swell of the newest septum, pointing upward to the preceding septum. In the radiographs, this vertical siphuncle was like a dipstick that showed us the decreasing water level. Experimentation with empty shells had shown that a chamber was half emptied when the water level had fallen below the entire siphuncle. It was at this point, then, that the nautilus moved for-

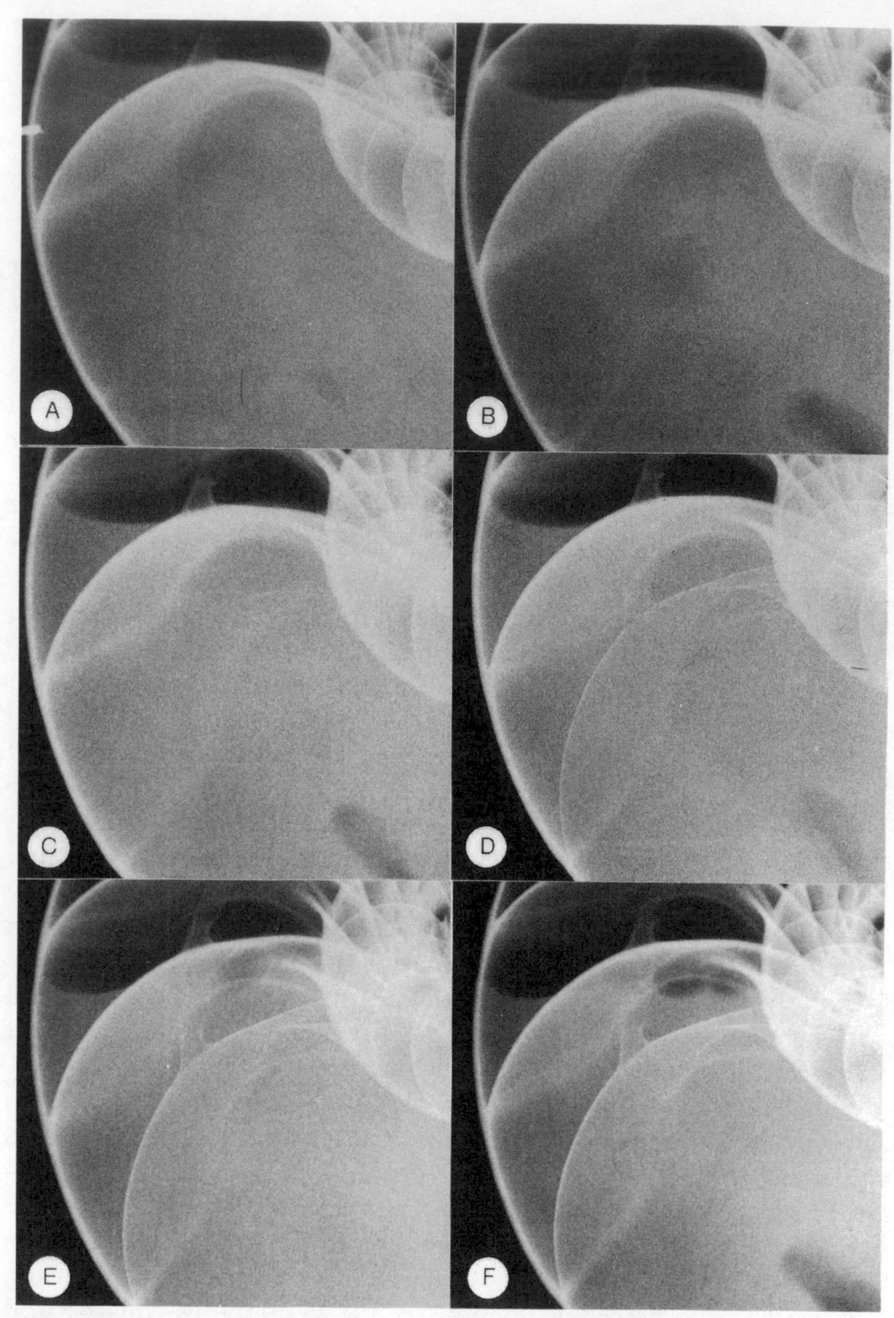

Radiographic view of new chamber formation. The six photos in this sequence show the formation and first emptying of a new chamber in a *Nautilus macromphalus* specimen at the Nouméa Aquarium.

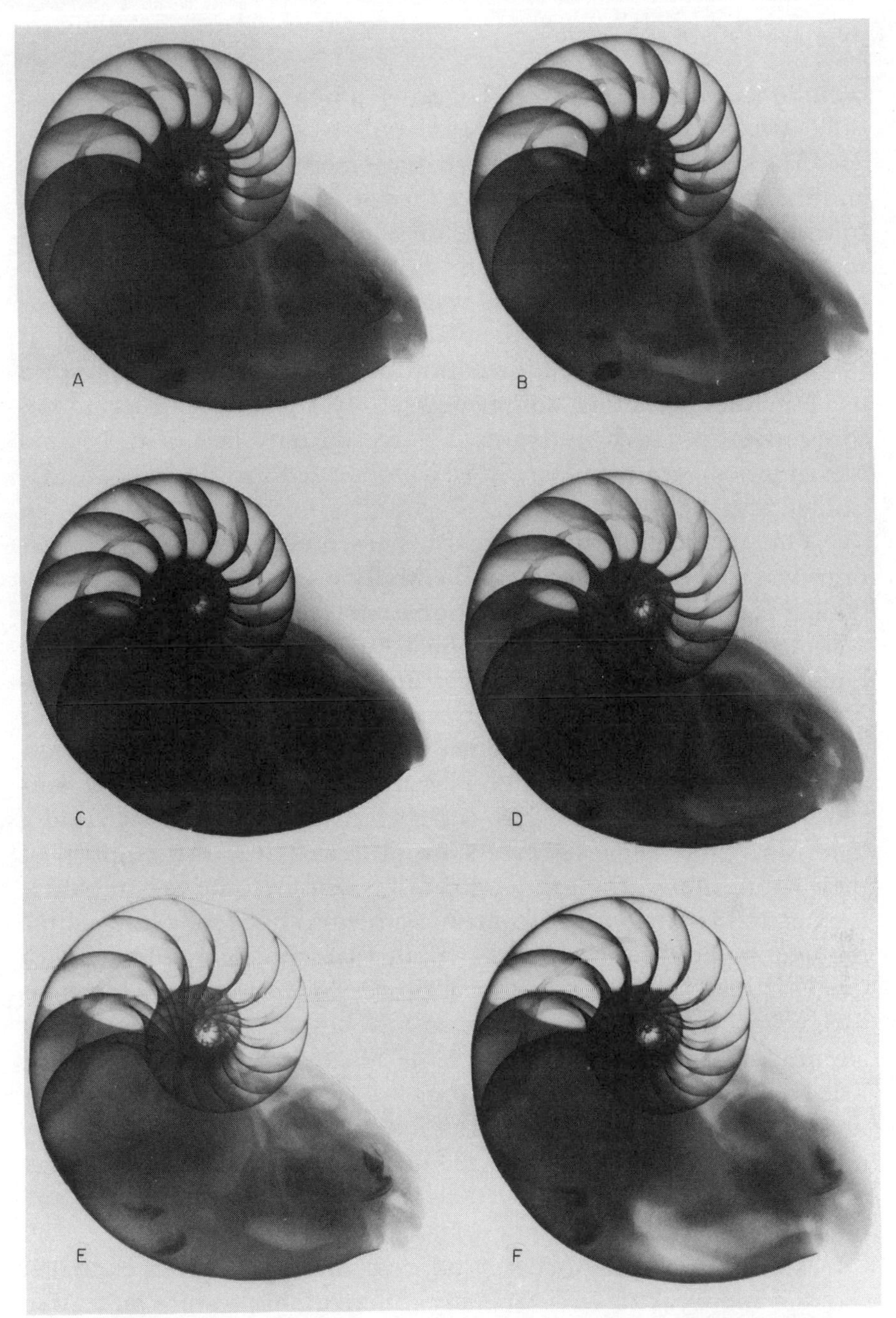

Liquid-emptying process in a specimen of *Nautilus macromphalus* over a four-week period. In each of the radiographs, note the enlarging bubble of gas visible in the last-formed chamber.

ward in its shell and began to lay down a new septum, thus closing off a new, completely liquid-filled chamber.

The correspondence between the emergence of the siphuncle and the initiation of new chamber formation was exceedingly useful to us. If we wanted to take tissue or shell samples from the earliest-formed stages of a new septum, all we had to do was watch the radiographs and the descending water levels to determine when new chamber formation would take place.

At first the chamber formation process seemed inadaptive to us. Why didn't nautilus completely empty a chamber of liquid before moving its body forward? The reason only half of this liquid was emptied became apparent as we observed the constant formation of new shell at the edge.

One of Denton and Gilpin-Brown's most important observations was that nautilus is always neutrally buoyant in the sea. Maintaining neutral buoyancy would not seem to be a major problem for completely grown nautiluses, which can simply remove sufficient liquid from their chambers to become virtually weightless in seawater and can compensate for the added weight of large meals or decreased weight from the sudden loss of shell material. For an immature nautilus, however, the rules are entirely different. As a nautilus grows, forming new flesh and shell material, its density rapidly increases. Unless new chambers are produced and then emptied of their liquid filling, the growing nautilus will become progressively heavier and less buoyant. To maintain neutral buoyancy, the animal removes liquid from chambers at a rate that compensates exactly for the flesh and shell growth. If a nautilus's weight in the sea increases by five grams in one week due to new growth, the animal must compensate by removing five grams of liquid from its chambers. The liquid within the chambers thus has two roles. First, it acts as a brace to support the thin new septum, which is slowly calcified by mantle tissue lining the back of the nautilus's body. Second, it serves as ballast that can be removed to compensate for new shell and flesh.

When a new chamber is produced, the new septum is at first too thin and weak to withstand any difference in pressure on its two sides. One side is subject to hydrostatic pressure of many atmospheres, transmitted through the animal's liquid-filled soft parts, while the other is subject to near-vacuum conditions that exist within the chamber as soon as emptying begins. This paramount fact was never understood by Lankester and Willey. A nautilus at a depth of 400 m has 40 atmospheres of pressure—almost 600 pounds per square inch—pushing against the new septum. If the emptying of the chamber liquid were to occur as soon as calcification of the new septum started, the hydrostatic force would easily crush the septum and, in so doing, kill the nautilus. The nautilus must wait a considerable period of time before it can dare to begin removing its chamber liquid.

Nautilus faces a paradox. It is constantly creating new shell and new flesh, but to compensate for this, it must continuously remove liquid from its chambers. Chambers, however, are produced at infrequent intervals, and during their formation there is a significant period when they *cannot* be emptied. Nautilus, then, must have sufficient reserves of liquid in its older chambers to take care of its buoyancy needs during the early septal secretion. It achieves this by beginning new chamber formation while the previous one is still half-filled with liquid. As Greenwald and I found in our 1979 studies, between three and four months may elapse between successive chamber formation events. An entire month of that period is needed to thicken the new septum enough to allow removal of chamber liquid behind it. During that month's time, the remaining liquid in the previous chamber or two is removed to take up the slack.

One unexpected result of our research was that we found that septa appear to be twice as thick as they need to be. Even more interesting was that nautilus can sense its depth and tailor its growth accordingly.

Greenwald and I wanted to know when the chamber would begin to empty. We assumed that the new septum would reach its final thickness before the liquid began to drain. But to our surprise, as we watched the septum gradually become thicker in successive radiographs, we saw that emptying started long before this. By measuring thicknesses of new septa directly from our X rays, we found that nautiluses in aquariums began to empty liquid while building new chambers when the septum had reached only one-third its eventual thickness. However, freshly caught nautiluses at the same stage of chamber formation emptied theirs when the septa were about two-thirds their final thickness. Nautiluses in the low-pressure environment of surface aquariums started emptying their chambers long before their deep-water counterparts.

The regularity of this process made us wonder. We conducted

experiments to demonstrate the controls as well as the processes of growth, and found that the amount of new apertural shell material played a key role in new septal formation. We discovered this when one of our specimens was inadvertently dropped during weighing. A large piece of shell cracked off the apertural region, and it took the creature several months to replace it. During this time liquid removal and new chamber formation ceased. When the front of the shell finally healed, the siphuncle resumed emptying the chamber liquid. Several weeks later a new chamber, now months late, began to form.

Due to decreased temperatures, the *Nautilus macromphalus* specimens maintained in the Nouméa Aquarium during 1979 and 1980 grew much more slowly than those studied by Arthur Martin in 1970, 1972, and 1975. Chamber formation at the cooler temperatures took between three and four months. At this rate a nautilus would complete growth in seven to ten years. This estimate, like all those previously made, was based on the assumption that each chamber took about the same amount of time to produce—in other words, that the rate was a constant. But was this assumption true? Again that wonderful element of scientific investigation, chance, stepped in. Up to this point, the only nautiluses we had caught were large, either completely grown or within one or two chambers of final size. In 1981, however, a shell collector in the Philippines discovered a way to catch large numbers of very small nautiluses and ship them alive to the United States. These small specimens were perfectly suited for studying growth and testing the hypothesis of constant chamber formation.

RELATIVE RATES OF CHAMBER FORMATION

The inability to trap small, juvenile nautiluses was one of the curious facets of nautilus studies. Small empty nautilus shells had been found on beaches, but no living nautilus of early post-hatching size had ever been captured in traps. Very small nautiluses had been observed on only one occasion. In the 1950s a shell collector was turning over rocks on a beach near Suva, Fiji, and under a particularly large one he came upon three baby nautiluses, all still alive. Although many investigators subsequently searched this area, they never found such small nautiluses here again. Because they were so rare, it was a surprise when a small fish importation firm in Los Angeles suddenly had large stocks of baby nautiluses for sale. The distributor knew nothing about their origin other than that they were from the Philippine Islands. To this day the area, environment, and depth where these tiny nautiluses were found, as well as the name of the exporter, remain a mystery.

The usefulness of these small nautiluses in answering fundamental questions about growth was immediately clear. When they were studied first in the New York Aquarium and later in the San Francisco Aquarium, the specimens were weighed and radiographed weekly or bimonthly. Some of the nautiluses must have been captured soon after hatching, because when first purchased, they had only 10 or 11 chambers and a shell diameter of slightly over an inch. In both New York and San Francisco, where these nautiluses were carefully fed and observed for over a year, it became apparent very quickly that the amount of time between chamber formation events increased with the addition of each new chamber. A very small nautilus could complete a new chamber in as little as two weeks. By the time it had reached near-adult size, this same nautilus took up to three months to complete one. In other words, as the nautilus reaches adult size, its chambers get bigger and therefore take longer to form. The assumption of constant chamber periodicity was false.

GROWTH RATES IN NATURE

The observations conducted in aquariums had revealed a wealth of information about the mode and tempo of nautilus growth. These studies had given the first estimates about the rate of development and indicated that the period between hatching and maturity could be as long as eight years. In the aquariums, conditions for growth were clearly far different from those in nature: The nautiluses were fed daily; they had more light; and perhaps most significant of all, water pressure was much lower. Nautiluses are generally trapped at depths of 300 to 1,200 feet, where the pressure ranges from 150 to 600 pounds per square inch. Nautiluses in aquariums live near the surface, at pressures of 15 pounds per square inch. Pressure differences have no effect on the growth rate of most marine animals. But in nautilus, pressure is decisive. And because pressure has such an effect on the emptying of chambers during growth, we expected that the depth where a nautilus lives would play a large role in determining growth rate: The greater the depth, the slower the growth. For us to finally determine the natural rate, we needed information about nautiluses in the sea, not in aquariums. But how would we do this?

Formation of two chambers in a single specimen of *Nautilus pompilius* over a five-month period.

The most obvious method was to catch and mark specimens, release them, and then recapture them at a later date, but this program would require extensive field work. To get around it, several scientists at Yale University began to wonder if geochemical information locked in the nautilus's shell might give us some needed information.

In the mid-1970s Karl Turekian, a professor at Yale, had developed a new method for dating mollusk shells. It has long been known that the radioactive decay of certain elements in nature can be used to date objects. A prime example is the carbon 14 technique, by which the age of organic objects thousands of years old can be determined. Most of these radioactive dating techniques are possible because the isotopes break down at a very slow rate. Carbon 14 isotopes take thousands of years to reach their half-life—the time it takes for half of the mother isotope to revert to the daughter isotope—while other radioactive tracers commonly used in geology take millions of years.

Turekian wanted to find some material whose half-life was in the hundreds of days, rather than thousands or millions of years. He found such a decay series in lead 210. Turekian knew that the ratios of naturally occurring radionuclides of lead could be found in the sea, and were incorporated into the shells of growing mollusks. He used this method to demonstrate natural growth rates of deep-sea clams.

In 1981 students and colleagues of Turekian published a report using this method to show rates of septal formation in nautilus. They analyzed two different shells. One specimen gave rates of about 30 days per chamber for six successively formed, larger septa. For a number of reasons, however, these results were thought to be minimum estimates. They got better results from a second specimen, which yielded figures of 60 to 90 days per chamber for eight successive chambers. When these investigators used the same approach several years later on freshly captured nautilus shells from Palau (a small island group in Micronesia), they found growth-rate figures of longer than 70 days per chamber. The radionuclide results indicated that growth rates in nature were as slow as, or slower than, our 1979 and 1980 aquarium-based observations.

NATURAL GROWTH RATES OBSERVED: SAUNDERS AND SPINOSA IN PALAU

By the late 1970s several lines of evidence suggested that nautilus growth rates were slower than anyone (except Oliver Wendell Holmes) had previously thought. Skeptics still wanted to see an actual growth rate for nautilus in its natural habitat. This was finally obtained during a tagging-recapture program conducted in Palau over a period of several years.

No nautilus had ever been found on Palau until the early 1970s, when Douglas Faulkner, *National Geographic*'s great underwater photographer, decided to try to trap for specimens there. Although nautilus shells were common on the beaches of Palau, no one had ever seen a live specimen on the island; it was thought that the shells had drifted in from the nearby Philippines. Faulkner wanted to find out if this folklore was true. He constructed large traps, attached them by rope to the side of the reef half a mile from shore, and waited. His early attempts were unsuccessful. He set his traps at 100 to 200 feet, the same depths at which nautiluses had been captured in New Caledonia and the Loyalty Islands, but he caught only fish and hermit crabs. On a final try, Faulkner went much deeper. Buying more line, he dropped one trap to 600 feet. As he pulled it up through the clear blue water, he saw the faint reflection of large white discs inside. To his amazement and delight, the trap contained several nautiluses; moreover, they were the largest living specimens ever seen. Faulkner had discovered a race of giant nautilus in Palau.

With mounting excitement Faulkner continued to trap, and soon found that these giant nautiluses were extremely common. The placid waters of Palau, unruffled by the raging trade winds that haunted investigators in other island groups, provided a perfect setting for large-scale nautilus research. Faulkner announced his discovery to the world, setting in motion events that would eventually lead to the first observations of nautilus growth rates in nature.

I first met Bruce Saunders and Claude Spinosa in 1976 while sipping champagne and eating nautilus canapés in the Brown Palace Hotel in Denver. The annual meeting of the Geological Society of

America was held there that year, bringing together geologists and paleontologists from all over the country for five days of scientific talks and hallway discussions. Those of us interested in cephalopods and paleobiology, calling ourselves Friends of the Cephalopods, had decided to inaugurate meetings at Denver and meet once or twice a year thereafter to discuss our research. To kick off our first meeting, I had arranged to bring in small chunks of frozen nautilus meat I'd obtained during my second summer of research in the Fiji Islands. At that time I had frozen some of the white meat from the adductor muscles and put it in a subzero freezer. Before the Denver meeting, I asked friends in Fiji to send the meat. They put it in a large thermos bottle packed with dry ice and airfreighted it to my Columbus, Ohio, address. This would have worked fine except that the package spent a weekend lost in Cleveland. When I finally received it and opened the thermos, the meat was cold but certainly not frozen. Hoping for the best, I repacked it and left for Denver.

The idea for a nautilus meal was born from stories of the paleontologist Roy Chapman Andrews that had intrigued me when I was a boy. One of my favorites was about a dinner of mastodon eaten by geologists at the turn of the century in Moscow. Siberian explorers had discovered a mastodon frozen in an ice crevice it had fallen into over ten thousand years ago. I had wondered what it must have been like for the assembled scientists to dine on this ancient creature. The Friends of the Cephalopods and I were soon to have a similar experience.

The affair at the Brown Palace was a night to remember. Early in the evening while the paleontologists gathered in that beautiful, venerable hotel, I was downstairs sparring with an ancient and infinitely displeased chef. This poor man had been asked to cook meat that was of dubious quality. "You have two choices," he said. "Run your ass over to Safeway and buy some scallops, or go for garlic." Just as I was about to take the Safeway alternative, I received an urgent call from the assembled greats impatiently waiting above. So the garlic would have to do. The small pieces of meat were quickly cooked on the grill and liberally smothered in garlic. Ten minutes later, champagne and the nautilus were brought into the ornate room on silver platters.

As the evening progressed and I was relieved of any fear that the nautilus meat had spoiled, I presented the results of the two expeditions I had conducted in search of nautilus in New Caledonia in 1975 and Fiji in 1976. After my talk, two men approached me to report that they planned to study nautilus in Palau the following summer. They were Claude Spinosa of the University of Idaho and Bruce Saunders of Bryn Mawr College. I asked them about their experience in small boats. "Minimal," they said. How about diving? "We are taking the course right now," they replied. Had they any experience in the tropics, or in Micronesia, or did they have a reliable guide who knew the waters? "None," they said. I promised to light a candle for them.

Spinosa and Saunders were both graduates of the University of Iowa, from a department known for paleontological studies of cephalopods. At Iowa Spinosa and Saunders, roommates and close friends, had come under the tutelage of two great students of the ammonoids, Professors W. Furnish and Brian Glenister. Furnish and Glenister were the foremost authorities on nautilus and the ammonoids, and had traveled to New Caledonia to view living nautiluses soon after Denton and Gilpin-Brown's research trip there. The two professors required their students to understand the biology of living cephalopods as well as their fossils, a rare practice among paleontologists of that time. To that end, students were expected to dissect nautilus bodies.

Late in the evening after classes, sipping a bit of medicinal Scotch, Spinosa and Saunders pondered the life of nautilus and began making plans about possible research they could pursue if they had a chance to study this exotic creature in its natural habitat. Years later, following Faulkner's discovery of living nautiluses in Palau, that chance presented itself.

In May 1977 Spinosa and Saunders arrived in Palau and found accommodations in a marine research station the Japanese government had built for the islanders. When they got there a great dispute over nautilus had just been settled. The director of the Palau marine research station at that time, Jim McVey, had been in the middle of a clamorous lawsuit over nautilus capture. The Palauans, who considered all of their marine life a national resource, were enraged to discover that McVey had set traps off their coast, even though after Faulkner's departure, he had caught only a few nautiluses. The Palauan legislature had taken up the matter, castigated the *hoeli* running their marine lab, and then promptly forgot all about it.

Avoiding fanfare as much as possible, Spinosa and Saunders borrowed McVey's single nautilus trap and set out on the lab's 16-foot boat, accompanied by a Palauan who claimed to know where nautiluses could be captured. After motoring around randomly for a while, the Palauan sheepishly raised his hands and disavowed any knowledge about the whereabouts of nautilus. The two investigators returned to shore with all of their capture gear still dry.

Based on advice Faulkner had given them before they left, Spinosa and Saunders began the slow trial-and-error process of learning capture localities and fishing techniques. They began to think about trap design and areas where nautiluses might be found. Their first solo trapping effort was a success. As they slowly raised the rebar (iron rods) and mesh trap from its 600-foot resting place, they saw that it was filled with nautiluses. All the frustrations of acquiring a boat, constructing traps, and dealing with Palauan politics and sensitivities were forgotten. The capture of so many specimens on the first effort boded well for the future.

But during the next ten sorties to sea, that promising future seemed somehow to come apart. The most they pulled in was a single specimen, and more often than not their trap was empty. The first bountiful capture seemed a fading memory. Then the team began to have enormous problems keeping their boat and equipment going.

After a rather complicated series of negotiations between the marine lab and a fisheries group in Palau, Spinosa and Saunders rented an open Boston whaler. The humid tropical climate was anything but kind to outboard motors and electronic equipment. Fortunately, whalers are built to float if swamped. Although the weather was generally calm, sudden squalls bringing heavy rain and wind periodically moved through the area. It was not uncommon to travel out under hot sunshine and then be overtaken by these rapid squalls.

To complicate matters, boat maintenance on Palau was strictly the province of a large Palauan named Abbie. Abbie had a lock on the local boat economy simply because he was the only reliable outboard mechanic on the island. Unfortunately he also had a typical Palauan sensitivity about deadlines. Abbie was much more concerned with fishing than with getting rich. The first time the two nautilus researchers brought their boat in for repairs, the promised next-day delivery began to stretch into weeks of waiting. Spinosa and Saunders didn't have the time to wait, and so learned to make small repairs with baling wire, rope, and hope.

Gradually their luck began to improve. Spinosa and Saunders had come with several scientific objectives but knew that their goals might change as the situation unfolded. As summer progressed, they began to catch numerous nautiluses and realized that they had an opportunity to conduct the first study of natural growth rates. They started tagging their trapped specimens with numbers and date of capture and releasing them back into the sea. It was not long before they recaptured one of their tagged specimens.

During the three-month period of their study, Spinosa and Saunders captured 375 nautiluses and tagged 247 of them. These large numbers are astonishing in comparison to what either Willey or Arthur Martin caught in their respective three-year research efforts. The calm conditions and deep-water areas adjacent to nearly vertical reef walls made capture off Palau easy and efficient. Also, in contrast to New Britain, New Caledonia, and the Loyalty Islands, buoy systems were not necessary; traps in Palau could be tied directly to the reef face and could never be lost. Of the 247 animals Spinosa and Saunders tagged, 31 were recaught during the 1977 expedition, and three were recaught twice.

The first piece of information to come out of this study was that nautiluses in Palau were found to swim long distances along the coastline. Spinosa and Saunders returned to Palau in 1978 and tagged another 460 specimens. The two scientists began to sail ever farther from their research base in an effort to determine the distances of nautilus travel. Tagged specimens began to appear in traps from all around Palau, even though they had been tagged in only one place. In a 1979 paper published in *Science,* Saunders and Spinosa reported that they had recorded Palauan nautiluses journeying up to 150 kilometers in less than a year, moving several kilometers some days and not at all on others.

Spinosa and Saunders saw a great deal of the open sea during their first summer. They spent virtually every day trapping, a practice that characterized the entire Palau research effort. They also had their share of close calls. On one occasion they decided on an experiment to test the hypothesis that nautilus maintains constant low chamber pressure, regardless of depth. They pulled up a trap to within 100 feet of the surface, and then dived down the line to the

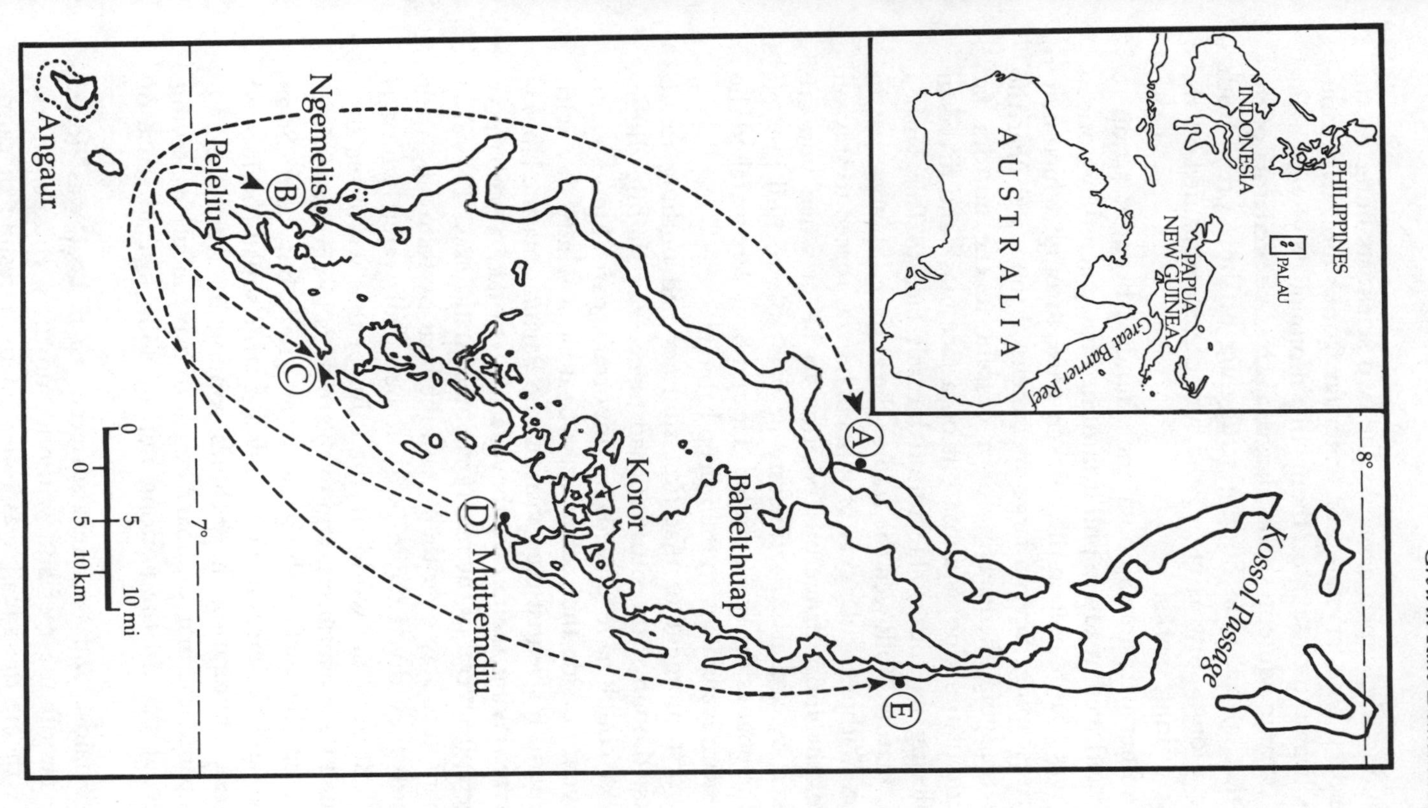

Map documenting the movement of tagged nautiluses in Palau. Of 247 animals tagged and released at sites B and D, 4 were recaptured 10 to 12 months later, approximately 25 to 95 miles away (A, C, and E). In 1978, 8 additional animals were recaptured at the same sites (B and D) where they had been released 5 to 11 months earlier.

trap to conduct the experiment. As the two scientists hung on the line in the deep blue, crystal-clear water far from shore, dark sinuous shapes began to close in on them. In a moment they were being circled by oceanic whitetip sharks, large, aggressive predators of the open sea that specialize in finding struggling offshore prey. Both men made a long, slow ascent to the surface, accompanied all the way by the waiting sharks.

On another occasion Spinosa and Saunders brought a trap to the surface and were ready to pull it into the boat when they noticed that there was a large hole in the chicken wire, through which nautiluses were escaping. Several large specimens in the water near the trap were readying themselves for a descent to the bottom far below. Without thinking, Spinosa put on a face mask and dived in after the fleeing nautiluses. He found himself face to face with a large shark that had followed the trap to the surface. Spinosa was in a tight spot. Without scuba gear or flippers he was forced to stay on the surface, the most dangerous position to be in when you are being stalked by a shark. Fortunately he was able to pull himself back into the boat as the shark circled. He and Saunders watched the nautiluses swim lazily back to their deep habitat.

The worst moment of that first summer had nothing to do with sharks, though if the situation had turned out slightly differently, sharks could have played an important part. Spinosa and Saunders were outside the reef in their boat on a windy day with another American scientist then visiting the Palauan marine laboratory. After retrieving their trap, they decided to take a scuba dive and photograph several of the newly captured nautiluses. They told their companion to stay over them with the boat by following their bubbles, then they dived down into the beautiful water, which is about the clearest in the world. In the calm of that warm sea they photographed their nautiluses, and in the process lost track of time. Meanwhile, on the surface things were anything but calm. Soon after Spinosa and Saunders had descended, a large squall had moved over the area, obliterating the bubbles coming up from the two divers. In a panic, the man on board began to drive in circles, trying to find the bubbles. In the pouring rain he soon lost all sense of direction.

When Spinosa and Saunders surfaced, their boat was gone. They were far off the reef and without buoyancy compensators. Happily for them, the squall was followed by a respite of calm.

They were just able to see their boat far in the distance, and by waving and shouting they finally attracted the driver's attention.

By their second season of fieldwork, Spinosa and Saunders realized that recapturing tagged nautiluses was going to be a frequent event. But for them to gain growth information, only incompletely grown nautiluses were useful—and only about 20 percent of those trapped were immature. The rest were all fully grown. Although the mature specimens could yield important information about movement, they revealed nothing about growth. Spinosa and Saunders hoped to overcome this problem through sheer numbers. They reasoned that if they trapped enough animals, they would eventually catch a juvenile they could measure for size at that time, then catch and measure again later to gauge its growth. They needed to recapture only one immature nautilus to confirm or refute the validity of the aquarium growth observations.

Claude Spinosa and Bruce Saunders returned to Palau in 1978 and 1982 to continue the trapping program; Saunders had been able to find enough money to go by himself in 1979 and 1981 as well. He was, to put it mildly, driven to recapture the elusive tagged juvenile specimen. In the absence of his partner, Saunders brought with him several student field assistants, who had jumped at the chance to visit a Micronesian paradise but soon learned the truth about their "vacation." Every day was spent in the boat, either setting or pulling traps, and then processing the animals: weighing, measuring, sexing, and marking each of the captured nautiluses. One of the principal capture sites was 40 miles from the marine lab's anchorage, a long ride in rough seas. The field assistants found themselves with a scientist who redefined the word *work*. Saunders cared little for days off or good food, and he was able to live weeks at a time on a diet of canned tuna fish and Snickers bars. When he expected his assistants to do likewise, several of them quit and went home.

Happily, Saunders found an assistant who was every bit as tough as he. Michael Weekley was a marine biologist who worked for the Waikiki Aquarium and was expert at diving, handling a small boat, and keeping nautiluses alive in aquariums. He was also a weight lifter and could stay alive on tuna fish and candy bars. The two of them began a collaboration that was tremendously profitable in terms of the scientific information gained.

As far as I can tell, both Saunders and Weekley were completely without fear. They would venture out in any weather and thought nothing of long hours, endless boat voyages, and spending an occasional night on the reef. They were also *very* hard on their equipment. The passage from the biological station to the outer reef was extremely complicated. It was a dredged pathway that zigzagged through shallow coral reefs. This channel was marked by reflectors rather than lights, and at night could be found only with flashlights. On clear nights this was relatively easy, but during squalls it was extremely tricky. Because of the frequency of their voyages, Saunders and Weekley encountered their share of bad weather. They were also disinclined to drive their boats at half speed. When I arrived in Palau in 1983, the Palauans gleefully pointed out "Saunders Rock," a coral block in the middle of the bay that Saunders had hit at high speed early in his career in Palau. Miraculously, he survived, but his fiberglass boat wasn't so lucky. (A later boat was dubbed *Yellow Submarine* for its propensity to swamp in high seas. The open boats rode so low in the water and were filled with so much gear that swamping was a common occurrence in the sudden squalls.)

At the time Saunders and Weekley were there, Palau was an extremely violent society, with one of the highest murder rates per capita in the world. Palau became a U.S. Trust Territory after American forces liberated it during World War II, and today Palauans live off tuna fishing and doles from the American government. Palauans have also developed a warm relationship with U.S. beer; Budweiser and Old Milwaukee are the two favorites and are consumed in unbelievable amounts.

Saunders and Weekley were evidently revered by the Palauans. Michael Weekley was their favorite *hoeli*, because he became more Palauan the longer he stayed. He was large and burly, built like a Palauan, and constantly had a rock-and-roll tune on his lips. He soon took up the habit of chewing betel nuts, the mild narcotic favored by the world's tropical peoples. The nut, which comes from a palm tree, is wrapped in a coconut leaf with a small piece of lime or coral. When chewed, the three ingredients yield both a soothing calm and a reddish saliva that finds its way onto most sidewalks, roads, and barroom floors. Chewing betel nuts eventually turns the gums red, while the added lime grinds the teeth down to points. When a beautiful Palauan woman suddenly smiles at you, revealing sharpened red fangs, the effect can be devastating.

WHERE ARE THE JUVENILES?

Saunders and Weekley stayed three to four months each field season. In addition to the nautiluses he and Claude Spinosa had already captured and released during their 1977 and 1978 seasons together, Saunders released 233 nautiluses in 1979, 452 in 1981, and, again with Spinosa, a whopping 978 in 1982. There comes a point in science, however, when generalizations are proven to everyone's satisfaction. Early in their work Saunders and Spinosa had admirably demonstrated that in Palau nautilus swims long distances and, unlike most other cephalopods, lives past its reproductive age. (An octopus, for example, dies soon after breeding. Nautilus apparently breeds for several years after reaching sexual maturity.) Why did Saunders continue the backbreaking work of capture and release? The answer is that even with the large number of recaptured nautiluses, very few showed any growth whatsoever. The greatest scientific prize for nautilus researchers would be the demonstration of actual growth rate in the wild. Although hundreds of juvenile nautiluses had been captured, tagged, and returned to the sea, none of the very small ones was ever recaptured, and out of 2,387 specimens originally tagged, only seven of those recaptured demonstrated measurable growth.

These seven specimens showing growth did yield interesting results. In four of them, the average rate of apertural shell growth was 0.10 mm per day, as compared to the 0.15 mm per day for aquarium nautiluses in New Caledonia. Two of the seven had finished growing. Assuming that the fastest observed growth rate (0.12 mm per day) occurred soon after hatching, Saunders estimated that maturity was attained at 14 to 17 years. As Saunders admitted, however, all of these results must be viewed with caution because they document only the last phase of growth, when the rate is known to diminish markedly.

Try as he might, Saunders never caught the elusive juvenile that would once and for all have solved the mystery of nautilus growth rates. His careful research, however, significantly increased our understanding of nautilus ecology, systematics, and growth. I believe that in his similar perseverance, single-mindedness, and resourcefulness, Bruce Saunders is our era's Arthur Willey.

A final and perhaps sad postscript of this story relates to the tagged but missing immature nautilus from Palau. I asked Claude Spinosa and Bruce Saunders why they thought they recaught so few juveniles. Spinosa thought that after capture the immature specimens moved to another area away from the traps, or refused to reenter them. Saunders thought that it was perhaps bad luck. My own feeling is that most were killed soon after being tagged. Although Saunders said that great care was taken to ensure that the captured nautiluses were kept cool during the long processing operation, they still would have been out of water or in small buckets for as long as an hour. In 1984 I attempted this procedure with nautiluses of various sizes. The mature specimens all survived, but most of the immature ones died, and the smaller their diameter, the quicker their death. The smallest, about half the size of a mature nautilus, were dead within a day. The larger specimens lingered for as long as four days before dying. The scavengers in Palau may have feasted after every nautilus catch and release.

Chapter 8

THE ROLE OF BUOYANCY CONTROL

Eric Denton and John Gilpin-Brown had originally demonstrated the principles of neutral buoyancy attainment and maintenance in nautilus. As we know, several major questions about the buoyancy system still remained unanswered: Is neutral buoyancy a passive system, or do nautiluses, like cuttlefish, change their buoyancy to suit their activities? One such use would be to change depth. Willey had proposed that nautilus migrates upward at night. Some scientists felt that such migrations could be powered only by changes in buoyancy: At the start of the night, a nautilus would rapidly remove some liquid in order to float upward into shallower water. At night's end it would submerge by adding liquid back into the shell.

Questions concerning rates of buoyancy change also remained unanswered. How quickly could a nautilus alter its buoyancy, if at all? If a nautilus eats a particularly large meal, does it compensate for its sudden increase in density by rapidly removing some liquid from its shell? Conversely, could a nautilus that lost shell in a predatory attack quickly add liquid back into its chambers to compensate for the sudden density reduction?

Anna Bidder's observations of nautiluses at the Nouméa Aquarium convinced her that rapid buoyancy change was possible. In *Nature* magazine in 1962 she wrote: "Changes in buoyancy were sometimes positive, sometimes negative, but the light animals would usually become heavy in a few hours." Eric Denton was not so sure. In a 1974 summary article dealing with his work on *Sepia* and nautilus he wrote: "It is not certain whether or not nautilus can adjust its buoyancy quickly." The primary goal of my research on nautilus—which had begun in 1975 and extended into 1982—was to determine how the creature used its buoyancy system.

THE EARLY EXPERIMENTS: LIQUID EMPTYING, 1975

When I arrived in New Caledonia in 1975, I implicitly accepted Denton and Gilpin-Brown's two-week estimate for new chamber formation. Because this estimate required that a new chamber be almost completely emptied of liquid during a two-week period, I assumed that a nautilus could remove at least several fluid ounces of chamber liquid each day. Such a rapid rate suggested that buoyancy-powered propulsion would be feasible. Arthur Martin and I so very much wanted to observe the process that we were guilty of a mistake common to many scientists: We *wanted* something to be true, and set out to make sure it was. We wanted nautilus to be capable of such rapid liquid removal. The adaptation seemed so elegant that we hoped to be able to demonstrate that it really did exist. We therefore designed a simple experiment that would demonstrate the capabilities of the liquid-emptying system.

We planned to drill holes in individual chambers of the shell, then put specific volumes of liquid in these chambers, seal them up, and after some period of time, measure the volume of remaining liquid. It sounded so simple in theory.

Drilling into the calcium carbonate shell was no problem. We practiced on empty shells using a power drill with a tiny bit. Our first problem came in trying to locate the position of individual chambers, whose limits are not marked in any way on the outside of the shell. More often than not we intersected two chambers at once. Unfortunately, we also badly mangled several animals by missing the chambers entirely, drilling instead into the soft parts.

Finding the positions of the chambers was simple, however, compared with resealing them. A nautilus shell is quite slippery, and we found that it resists most glues and sealing compounds. We had to find a material that would set quickly and was nonlethal to the nautilus. Also, because our specimens lived in aquariums for days or weeks after the start of our experiments, we had to find something that would adhere underwater for these protracted lengths of time. We tried plugs and many kinds of cements. Our scavenger hunts through New Caledonia's grocery and hardware stores in search of

Two nautiluses undergoing buoyancy experiments in an aquarium. Corks in the shell seal holes drilled into individual chambers.

new sealant technologies were hilarious. Art Martin spoke no French, and even though I did, my accent left no doubt about my country of origin and a great deal of doubt about what I was trying to say.

When we eventually found a fast-setting dental epoxy that lessened our fears of leakage, we set out to quantify the rates of chamber liquid removal. Unfortunately, we couldn't believe our results. We repeated experiments over and over, but each time reached the same conclusion: Either our seals were constantly leaking, or the rate at which liquid was emptying from the chambers was far slower than the period described by Denton and Gilpin-Brown. We assumed that our seals leaked.

As the weeks progressed, the fastest rate of liquid removal we observed continued to be about one fluid ounce per chamber per day, and in most cases it was even slower than this. Martin and I realized that the process was far less rapid then we had hoped and expected. We also arrived at another interesting conclusion: The greater the volume of liquid put in at the start of the experiment, the faster it would empty. The rate of liquid removal, therefore, seemed to be affected by the nautilus's initial buoyancy. If we added a small volume through the hole drilled into the shell, the rate of removal was low. The highest rates of liquid removal occurred in specimens whose shells had been drilled into in several chambers and then filled with liquid. Adding various weights to the shell could also influence liquid-emptying rates. It was clear, then, that a nautilus could sense its weight in seawater and tailor its liquid-emptying rate accordingly.

A second question that concerned us dealt with the effect varying depth had on liquid emptying. Denton and Gilpin-Brown's observations on cuttlefish showed that increasing depth made osmotic liquid removal more difficult. We therefore decided to put our nautiluses back into the sea in closed cages at depths of 60 to 120 feet. We repeated our liquid-emptying experiments at various depths for at least a week. As we expected, rates of liquid removal became slower with increasing depth.

We wished to test one of the most important remaining questions concerning nautilus buoyancy: Did these creatures use buoyancy change to power the nightly excursions into shallow water they were assumed to make? All of the newly captured specimens from our traps had been weighed in seawater on a torsion balance, and our results confirmed Denton and Gilpin-Brown's findings: Freshly captured nautiluses have a slight weight (negative buoyancy) in seawater. But would a nautilus captured at night, when presumably it would just have risen from the depths, show the same result? Unfortunately such specimens could not be caught with traps. So to test our hypothesis, we would have to capture nautiluses at night by hand.

DIVING FOR NAUTILUS, 1975

Believing the reports lobster fishermen made of nocturnal sightings, French oceanographers stationed in Nouméa in 1975 thought they could find nautilus in shallow water at night. Also, Jacques Cousteau and his crew had shot television footage of the creature at night, although there was some speculation that this film had been staged. I very much wanted to obtain some shallow-water specimens and measure their buoyancy as soon after capture as possible.

The marine scientists at the ORSTOM laboratories were of great help in this project. Before our first attempt in September 1975, I madly rushed about town in the late afternoon for film and batteries for my underwater light. We set out from Nouméa in a Boston whaler crammed with the piles of gear necessary for four divers. Smashing through the chop in the lagoon, we made a mockery with our speed of the seemingly endless trips I had endured in the old fishing boat with Arthur Martin. We anchored outside the barrier reef by the last light of the setting sun. I rather nervously watched large dorsal fins pass lazily by as my three companions helped themselves to paté, bread, Camembert, and liberal doses of red wine.

We waited until 7:30 to don our gear. The large double tanks my colleagues favored were awkward to maneuver in the boat, so we lowered them into the water on lines and put them on underwater. Rigging up the "chief," our head diver, Pierre Laboute, was especially time-consuming. Laboute, who also carried two cameras and flash units, wore a motorcycle helmet with a light affixed to it. In the complete blackness I took a last look at the distant, twinkling lights of Nouméa, 12 miles away, and then rolled into the sea.

We put a flashing strobe on the bottom of our boat to mark our return path so that no one would surface after a long dive and then have to search for the vessel. We dropped anchor just offshore of the reef flat. From this breaker zone, the reef descends gradually downward to a depth of about 40 or 50 feet. It is within this zone that coral growth is greatest, creating a riot of shapes and colors during daytime hours; but by dark only branching silhouettes were visible in our light beams. At night the reef comes alive with swarms of invertebrates and larger plankton that lie concealed on the bottom during the dangerous daylight hours. Lobsters steal out from their

Pierre Laboute readying gear for a night dive, New Caledonia, 1975.

holes, as do octopuses and many rare gastropods. Fish, on the other hand, are largely inactive.

Laboute and I dived as quickly as possible to get out of the surge caused by long rollers of the Coral Sea. We then followed one of the large coral gullies that crosses the reef top, finally coming to the edge of the reef. We knew we were coming up to a sheer vertical cliff that dropped into the depths. With our lights we probed the blackness of this dark abyss. Far away, a pair of tiny, dim lights gave the positions of our two companions. Laboute swam out over the precipice, disappearing from my light beam as he dropped slowly downward. I followed him and began my own slow fall toward the bottom of the Pacific Ocean.

We stopped at 150 feet and clung to the edge of the reef wall.

Laboute turned off his light and motioned for me to do the same. We were instantly surrounded by the velvet blackness. But after a minute, our eyes adjusted to night vision and the swirling, photoluminescent plankton turned the sea around us into a fairyland of light. The slightest movement of our fins provoked thousands of tiny organisms to fire their photons in protest. Laboute's light snapped back on and the spell was broken. He began to kick his way upward; the needles on our decompression meters showed our steady uptake of nitrogen, and neither of us wanted to spend any time on a line, decompressing in the nighttime sea.

We slowly swam up the face of the cliff wall, stopping for an occasional picture. Each time, the detonation of our powerful camera strobes would light up the entire area. Laboute pointed out sleeping parrot fish, their bodies enclosed in masses of mucus. He deftly captured a lobster or two. And when we finally came over the top of that dark cliff, his light transfixed a faint white sphere far to our right. We had found our first nautilus.

Laboute swam along the reef edge toward the nautilus, his light illuminating the white shell. When we first saw the creature, it had apparently just settled onto its prey, a large spiny lobster. As we approached, the nautilus jetted off the bottom in flight, carrying the lobster with it. Laboute and I pursued and soon caught up. As we photographed this nautilus with its prize, I began to understand for the first time how completely we, the so-called specialists, had misjudged our animal. For years scientists had thought buoyancy change was the key to nautilus's nightly migrations, and that this poor, frail creature was too weak to swim upward for hundreds or thousands of feet. Our specimen had just shown us that it could move vertically through the depths quite easily by swimming rather than by changing its buoyancy.

On returning to the boat with our prize we had two surprises. We found that our nautilus had captured a freshly molted exoskeleton of a lobster, not the lobster itself. Even when we put our nautilus into a bag, it refused to release its grip on the exoskeleton and continued to eat away at it. We also found that our two companions had captured a nautilus as well. Within an hour we were inside the Nouméa Aquarium, attaching the two specimens to a torsion balance. Both gave buoyancy measurements similar to values for trap-caught specimens. Neither showed any evidence of positive buoyancy.

Nautilus macromphalus carrying lobster molt, New Caledonia, 1975.

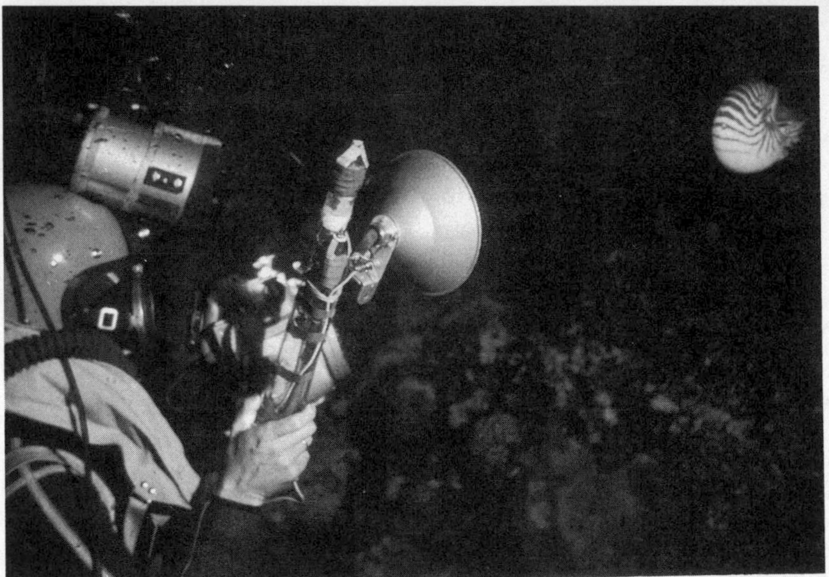

Pierre Laboute photographing a *Nautilus macromphalus,* New Caledonia, September 1975.

During the following weeks, and again in 1977 and 1978, I continued to dive at night for nautiluses. These dives were always extraordinary experiences. On one such occasion with Laboute I followed a nautilus into very shallow water, where crashing breakers knocked it against the walls of the reef. This nautilus was making a beeline for a crab molt and seemed nonplussed by the heavy wave action. On another occasion a small shark swam between us. We could see only its eyes, which shone, like a cat's, an intense yellow-green in our lights.

In 1978 I took a night dive with Pierre Djouman, the Nouméa Aquarium's chief diver. After a particularly long, deep dive we had to decompress on a line at ten feet. Pierre began flashing his light outward into the darkness, nervously flicking its beam in many directions. Suddenly he bolted for the surface, several minutes before we could ascend safely, leaving me hanging on the line. When I rejoined him in the boat he couldn't believe that I had not seen the large tiger shark circling us. Two weeks after this episode Pierre was killed while diving for a nautilus trap in the same place. He made it to the surface, but was badly embolized. The wet suit of his partner, who was never found, was later discovered in the stomach of a large shark.

The dozen nautiluses caught by hand on these dives yielded very interesting information. All were sacrificed soon after capture. I preserved the stomach contents, which were later identified by crustacean specialists at the Allen Hancock Foundation in Los Angeles. Their analysis confirmed that *Nautilus macromphalus* in New Caledonia eat molts of many crustacean species and seem to prey on live specimens of a single species of hermit crab as well.

LOCAL OSMOSIS

Discovering the osmotic mechanism nautilus and *Sepia* use to empty their chambers of liquid was one of Denton and Gilpin-Brown's major achievements. The two scientists demonstrated that chamber liquid is made less saline than the animal's blood, and that once that difference in salt concentration is achieved, the chamber liquid will cross the permeable membrane of the siphuncle, flow out of the chamber, and enter the bloodstream. However, emptying will take place only if the osmotic pressure created by the difference in the salt concentration of the two fluids is greater than the ambient hydrostatic pressure, which is determined by depth.

Denton and Gilpin-Brown realized that if liquid emptying was to continue, this difference between the salinity of chamber liquid and blood would have to increase as the animal descended deeper into the sea. For instance, a cuttlefish living at a depth of 300 feet would have to remove far more salt from its chamber liquid than it would at 100 feet. Following this reasoning, Denton and Gilpin-Brown realized that there would be a definite bottom limit to this system. The water within the chambers is secreted by tissue, which means it initially has a salinity equal to that of blood (and seawater as well). This liquid contains a finite amount of salt. Theoretically, if all of the salt ions were removed from it, the chamber liquid would be essentially fresh water, but in reality this never happens; the liquid always retains salt in some concentration that keeps it from being defined as fresh. Therefore, the maximum differential that could exist in this system would be between the salt levels in fresh water and in blood (or seawater). The osmotic pressure created by that ratio equals the hydrostatic pressure that exists at a depth of 240 meters, or about 800 feet. This is the maximum depth at which the chambers could be emptied. But we knew nautilus could be trapped far deeper than 800 feet, so was this a reasonable maximum depth limit?

Cuttlefish are creatures of very shallow water. Because the species common to Europe is rarely found any deeper than 200 to 300 feet, it does not have the problem of the 240-meter osmotic depth limit. The same cannot be said of nautilus. Although the nautiluses Denton and Gilpin-Brown captured in Lifou were taken at depths of only about 300 feet, they knew that others had been captured as deep as 1,800 feet, far deeper than the 800-foot limit scientists had supposed. Denton and Gilpin-Brown were confronted with two possibilities: Either the maximum depth information about nautilus was wrong, or the creature could descend to depths greater than 800 feet. If the latter was true, then the nautilus was using some liquid-removal mechanism other than the one the two scientists had proposed.

In the 1970s it had become clear that nautilus could be caught deeper than 800 feet. During my visits to Fiji I had captured specimens of *Nautilus pompilius* at depths of 800 to 1,600 feet, and most commonly at 1,000 feet and deeper. I had firsthand evidence that proved nautiluses did not live only in depths of 800 feet or less. Either they were descending further for only short periods of time, or their liquid-emptying system was different from what Denton and Gilpin-Brown had envisioned in their 1966 paper.

In another paper, published in 1974, these same two scientists proposed an alternative to their 1966 model of "simple osmosis." They noted that the osmotic system's driving force was the difference in the salt concentrations of chamber liquid and blood. The limiting factor of this system was the amount of salt that could be removed from the chamber liquid. Once most of the salt was removed, the maximum concentration difference was achieved. The only alternative way to create an even greater difference was to make the blood saltier! Even though this is impossible for the entire blood volume, what if only tiny amounts were made saltier? What if a nautilus could establish higher concentrations of salt in blood contained entirely within the siphuncular tissue itself? If this happened it would produce an osmotic differential far higher than what could occur between normal blood and fresh water. This was the possibility Denton and Gilpin-Brown raised in 1974. Moreover, there was precedent for this proposal; a similar system had already been identified in various mammalian organs, such as the gall bladder. This type of emptying, produced by locally high salt concentrations within tissues, was called *local osmosis*.

Denton and Gilpin-Brown proposed that nautilus used such a system. Anything that would make chamber emptying more rapid and efficient would clearly be advantageous, even at depths of less than 800 feet. A nautilus must not only empty a chamber, but must also keep it that way for the rest of its life. These are really two separate problems. If a nautilus emptied a chamber, then "switched off" the siphuncular pump, liquid would be forced slowly out of the blood and refill the chamber, and soon the creature would be too heavy to swim. The siphuncular tube could conceivably be blocked with calcium carbonate cement after a chamber was emptied to prevent liquid from reentering. But nautiluses do not do this. An adult nautilus can have up to 36 chambers, and in every one the siphuncle is functional. The animal must keep the pump operating in each of these chambers throughout its long life. Chambers are first emptied of liquid, a process lasting weeks or months, and then they are kept free of liquid by the osmotic pump.

THE SALT ENHANCEMENT EXPERIMENTS

In theory, the local osmosis model for emptying chamber liquid seemed easy to test. All one had to do was place a nautilus in a sealed cage at a depth greater than 800 feet and leave it for some length of time. Upon recovery, if the chamber liquid was equal to or less than the amount present at the start of the experiment, local osmosis had occurred. In reality, however, carrying out this experiment was anything but easy.

In 1979 Lewis Greenwald, the Ohio State physiologist, and I started working on the various alternatives. Lewis hit upon an ingenious test of the local osmosis hypothesis that did not require experiments in the sea. We could reproduce conditions presumably occurring within the chambers of a nautilus at depths greater than 800 feet by replacing the normal chamber liquid with saline solutions. The salt concentrations in this artificial chamber liquid would be *higher* than those of either seawater or blood. A nautilus could remove such a solution only by creating, *within the siphuncular tissue,* a local concentration of salt even higher than that of the liquid being emptied. This reaction is characteristic of local osmosis systems.

In July that year we began these experiments in the Nouméa Aquarium. We purchased a small, portable radiograph machine from a veterinary supply company that enabled us to radiograph our nautiluses right next to their cooled seawater tank. The radiograph soon became an indispensable tool in our buoyancy studies, because it let us see into the shell of a living specimen and correctly choose our target chambers.

Greenwald and I began the experiments using solutions that were slightly more salty than seawater, then progressed to solutions with about twice the salt concentration of seawater. These were injected back into the chambers through a quarter-inch hole we had drilled through the shell. We then sealed the holes with specially made rubber corks. After an anxious week we uncorked the chambers and carefully decanted the liquid. To our delight, there was a significant reduction in liquid volume. We had demonstrated that emptying of such a solution could occur. We wanted to perform one further test on this liquid, however; we knew that nautiluses readily removed salt ions from chamber liquid, but if our experimental animals had removed enough salt from the artificial chamber liquid to make it *less* salty than seawater, our test would be invalid. Lewis Greenwald placed a sample of the chamber liquid into a vapor-pressure osmometer, the laboratory tool commonly used to measure salt concentrations of liquids. The osmometer began its run, streaming meaningless numbers as it pondered the solution we had inserted. As it finished, it gave a loud chirp. We stared at the readout. The sample's salinity was still far higher than that of seawater. A nautilus could empty solutions saltier than seawater. Local osmosis was demonstrated.

We repeated this experiment on many different specimens, gradually increasing the starting concentration of our test solutions, and found that nautiluses could empty solutions nearly twice as salty as normal seawater. One surprise, however, was how slowly they accomplished this. Though nautiluses could seemingly empty their chambers or keep them that way at depths greater than 800 feet, the emptying rates were astonishingly low.

Greenwald and I wanted to see if local osmosis worked in nature as well. We put nautiluses in closed traps at various depths. Most of these experiments failed; the specimens either died in the cages or, more often, we lost cages and their inhabitants. Because the rates of emptying demonstrated in our salt-enhanced experiments proved so slow, we knew that one to two weeks' time in the sea was necessary for significant emptying to take place. The prospects of recovering a cage two weeks after it had been put in the sea in New Caledonia were poor.

Gradually, however, we recovered enough of these experiments to get a picture of emptying rates in nature. Emptying was found to occur at depths of about 300 meters, or 1,000 feet. At greater depths the nautiluses always had more water in their chambers than they did at the start of the experiment. We also found that when flooding occurred, it took place in all the chambers, not just the last two or three. Nautiluses can descend to depths of about 1,800 to 2,000 feet before their shells implode. Although their average habitat depth is probably less than 1,000 feet, they can, and do, dive deeper. However, they can stay at these depths only for limited periods of time; otherwise their shells slowly fill with water.

One of the great puzzles about nautilus concerns its slow growth rates in nature. Based on Saunders's results from Palau, it appeared that a nautilus might take as much as 15 years to reach final size. In the aquarium, however, growth was much faster, even at temperatures equal to those in nature. Perhaps the steady food supply in captivity accounted for more rapid growth. But perhaps the pressure differences between the two environments was even more important. Could the slow liquid-emptying rates somehow be related to growth rate?

A striking finding of the liquid-emptying experiments was that the emptying rates dropped as depth increased. At the surface, a nautilus removes about one fluid ounce of liquid from a chamber in a 24-hour period. At 300 feet, the same nautilus removes only one tenth of that volume in the same period. At ever-greater depth, the emptying rates decrease. And because growth can continue only as long as new chambers are created (and then emptied of liquid), it appears that liquid-emptying rates create stringent limits and controls on development. Our experiments suggested that a nautilus grows as fast as biology will allow, and that the the liquid-removal rate is the factor that limits the rate at which this can occur.

NEW CALEDONIA, DECEMBER 1982

We had been at sea for three days and were slowly moving northward up the New Caledonia coast. The heavy following sea was making all of us queasy, and I was already dreading our return voyage, when we would have to fight *against* the huge rollers. It was typhoon season, and a hurricane had run down the slot between the New Hebrides and New Caledonia, making a loop around the southern part of the island. It sat brooding roughly a hundred miles to the south of us, making the air heavy, windy, and wet.

Life on board the ship was just as turbulent. I was a visiting scientist on the French oceanographic vessel *Vauban*, an 85-foot-long study in rust and age. In the early 1960s she had escaped Madagascar's revolutionary uprising. When that French colony threw out its overseers, the *Vauban*'s captain slipped her to sea with a skeleton crew and many scientific and political documents, earning himself a prison sentence in absentia from the revolutionary Malgache committee. The captain radioed French authorities, who instructed him to make for New Caledonia. The old vessel, built and used for coastal surveys, not the high seas, underwent an incredible three-month voyage before finally arriving in New Caledonia. The captain's seamanship and authority were unquestioned. In 1982, in his second decade as master of the *Vauban,* he wondered if he would have to flee yet another French colony. The Kanak independence movement was in full cry, and that cry was for blood.

Out on the heavy seas we had revolutionary history very much on our minds. We were now near the northern tip of New Caledonia, and although we had never left the sight of land, we had been passing deserted coastline for two days. On land things were definitely heating up. The austral summer was in full season, and the oppressive humidity and impending typhoon had put the island on edge.

A white Frenchman had built a sawmill in a rural region of New Caledonia. He needed wood and was given permission to log an area considered sacred by the Kanak tribal council that represented the indigenous New Caledonians. In protest the Kanaks barricaded the main road leading to the logging camp. For two months the barricades were in place while both sides negotiated. The French finally ran out of patience and five white gendarmes were ordered to drive toward the barricade. Several miles from their destination they drove through a narrow canyon and into an ambush. Bullets from above killed three of the gendarmes instantly and left the other two seriously wounded. Two of the dead men were in their early twenties and had been in New Caledonia less than two weeks.

The news of the ambush and massacre came over the radio on our second day at sea. The *Vauban* had a crew of four officers and engineers, all white, and 12 seamen, all Kanak. The officers ate in a separate mess and had wine, whiskey, and fine French cuisine. The crew was forbidden alcohol, was served different food, and lived, ate, and slept apart from the officers. After the announcement of the massacre, we looked at one another, officers, scientists, and crew, as the radio droned on about reprisals and countermeasures.

The *Vauban*'s mission on this voyage was to try out various deep-water fishing techniques. We carried a massive new type of nautilus trap, a huge cube of rebar (iron rods) and wire six feet on a side, its base lined with several hundred pounds of iron chain. We also had generators and portable radiograph equipment that allowed me to x-ray newly caught nautiluses within minutes of capture. Our catches had been very successful; I had over 20 living specimens swimming in the ship's great hold. On our third day we were approaching the site of a trap put into the sea the night before.

We knew there were problems when we saw the huge buoy marking the trap's location a mile from the point where we had originally dropped this giant cube. After snaring the buoy and attaching the thick cable to the ship's winch, we began the slow haul that would raise the trap we hoped would be filled with nautiluses from its 400-foot depth. As the giant winch pulled in its load, the line began to move laterally to the ship, *as if a large fish were playing it,* I thought to myself. Suddenly, the seamen staring down into the clear water shouted to slow the lift. As the trap came to the surface the water started to boil from the thrashing of a large gray tail wrapped in line. A huge tiger shark was slowly pulled onto the deck, very much alive, and as far as I could tell, quite unhappy.

It was a madhouse. The shark, now free of the confining cable, thrashed violently on the rolling deck, upsetting rope and ripping cables as it writhed and snapped massive jaws at its tormentors. The first mate ran for a gun. I was trying all at once to take pictures, stay out of everyone's way, and keep my limbs safely attached to my body. A huge Kanak from Lifou, a member of a champion rugby team and an ardent proponent of independence from the French, appeared from somewhere with a fire ax. He arrived at the same time as the mate, who held a pump-action 12-gauge shotgun. In one of those clear, unending moments, the armed white man and the armed black man stared at each other from either side of the now forgotten shark. Suddenly the black man swung his ax, severing the shark's backbone. When I later measured the remains of this great fish, the monster I expected to be at least 20 feet long was a mere 14 feet.

The trap was mangled. The tiger shark, attracted by the bait inside, had chewed and distorted the rebar and iron to get its meal. In the process it had wrapped itself in the strong cable leading to the surface buoy and eventually dragged the entire rig for a mile. With the trap beyond repair and his crew on edge, the captain decided to head back south for Nouméa.

The old ship turned into the heavy sea and started its three-day fight against the wind. She began to pitch and crash into the waves. The giant shark was left on the deck and by afternoon had rotted, creating the most horrible stench imaginable. I was sick for three days as we slowly bashed southward through the waves. On land the gendarmes' assassins were captured, jailed for six months, then freed.

EXPERIMENTS ON COMPENSATORY BUOYANCY CONTROL

As happy as I was to return to dry land at the end of that nightmare voyage, I was happier still to finally have healthy *Nautilus macromphalus* swimming in the cooled tanks of the Nouméa Aquarium. These animals were to be used in my final experiments on buoyancy, a subject I had begun to study during my first visit to New Caledonia in 1975. The goal of all of these experiments was the same: I wanted to know the uses and limitations of the nautilus buoyancy system.

In 1982 I was ready to ask a new question: Could short-term, *compensatory* buoyancy change be effected? Up to this point, most buoyancy experiments had been related to questions of shell growth. Removal of chamber liquid was known to compensate for the long-term process of growth, but could the nautilus's buoyancy system have other uses as well? Could buoyancy change counteract the added density of a particularly large meal or the reduction of density caused by sudden shell loss?

The latter of these two questions seemed the more interesting. As early as 1975 I had experimented to see if liquid could reenter a chamber after that chamber had been emptied normally. I attached floats and buoyant corks to various nautilus specimens to see if they would compensate for their suddenly acquired positive buoyancy by refilling their chambers. I found no evidence of reflooding. I had not been convinced that these experiments were valid, however. The specimens I studied had already been used in other types of experiments, and none were freshly captured.

The situation changed in 1980 when Lewis Greenwald and I demonstrated that freshly captured nautiluses *would* partially refill their chambers after sudden buoyancy change. When we made a nautilus more buoyant by removing all its chamber liquid or by removing shell material from the aperture we observed liquid slowly reentering the specimen's chambers.

By 1982 I had all of the tools necessary to answer my questions satisfactorily. On the voyage of the *Vauban*, I had obtained many healthy fresh specimens, and the Nouméa Aquarium's two nautilus tanks kept them at about 55° F. The radiograph and drilling proce-

Nautilus with large but rehealed shell break. This type of damage would have required a compensatory buoyancy response.

dures were well tested. And I had a new tool to complete the instrumentation: Brainweigh.

Brainweigh was a newly purchased electronic balance modified to sit atop an aquarium and weigh nautilus in seawater. Before Brainweigh, measuring the weight of a nautilus in seawater—the necessary step in establishing both its buoyancy and its density—was a tricky procedure and could be accomplished only with anesthetized or dead specimens. Living, awake specimens thrashed about on the balance; and urethane, the anesthetic we used, worked well for single measurements but was highly carcinogenic to humans and killed nautiluses after five or six measurements. I

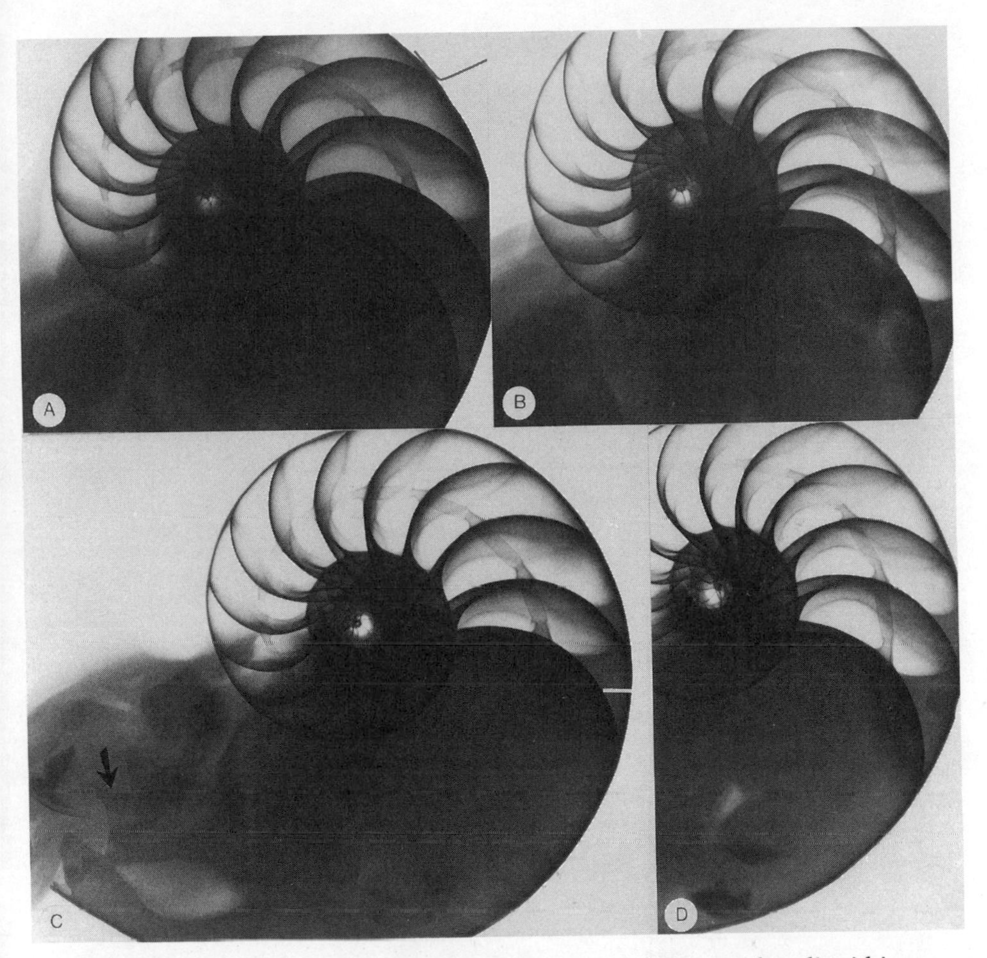

Radiograph sequence demonstrating nautilus's ability to replace liquid into its chambers. In figure A, the specimen's last chamber is completely filled with liquid. Immediately after this radiography was made, the shell was drilled and all liquid removed. In figure B, the nautilus has refilled the empty chamber.

wanted a system for weighing a nautilus in seawater every couple of minutes, without resorting to anesthetics.

I found that I could obtain this measurement accurately by tightly wrapping the specimen in cheesecloth and putting it into a plastic box filled with seawater. First the box was suspended from Brainweigh and the balance zeroed. Next the nautilus was quickly

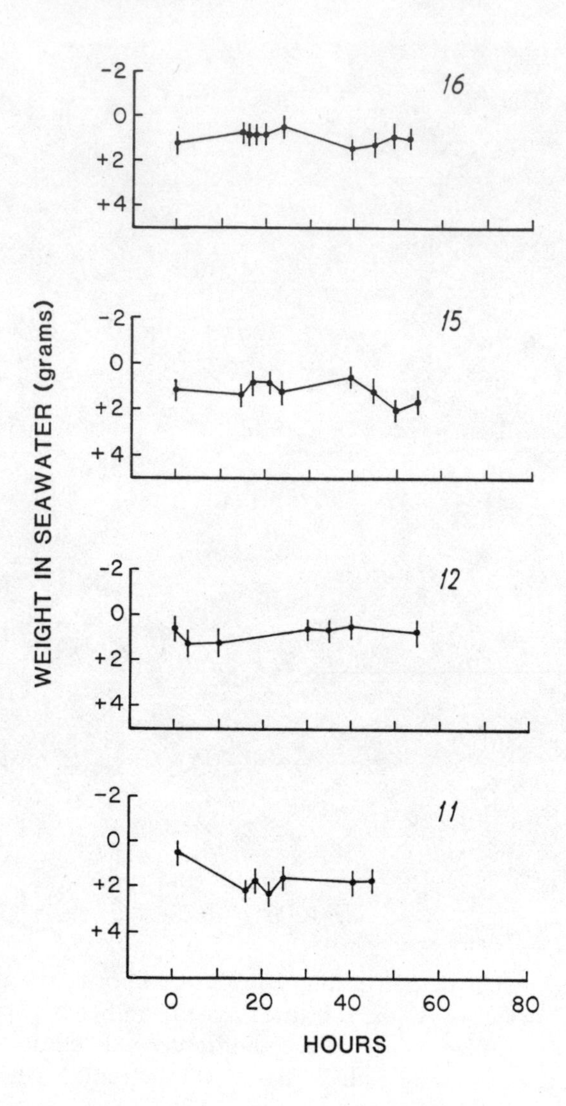

Figure showing normal patterns of buoyancy in four different nautilus specimens. None of these animals was manipulated in terms of its buoyancy; these measures can therefore serve as controls against which the following experiments involving sudden buoyancy change can be compared. The vertical bars on the graphs refer to estimated experimental error (0.3g). Experimental error comes from sensitivity of the balance (0.1g) and the weighings themselves. Although no subsequent graphs show the error bars, all points listed on subsequent graphs have similar estimated error ranges.

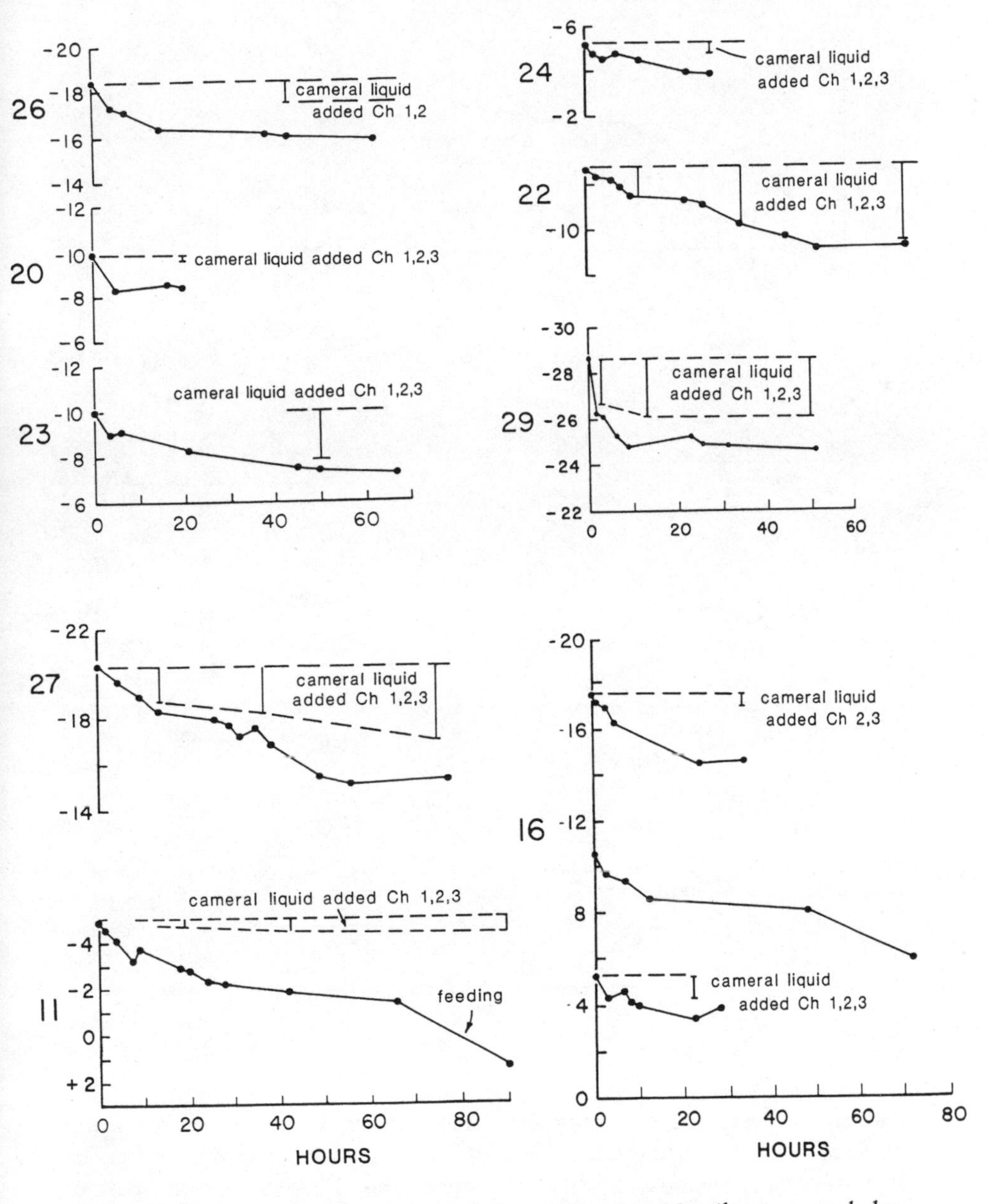

Patterns of compensatory buoyancy change in nine *Nautilus macromphalus*. The vertical axis in each graph refers to weight in seawater; the horizontal axes show the number of hours after the initiation of each experiment. The portion of each graph labeled "cameral liquid added" shows the amount of buoyancy change that can be attributed to measurable chamber refilling.

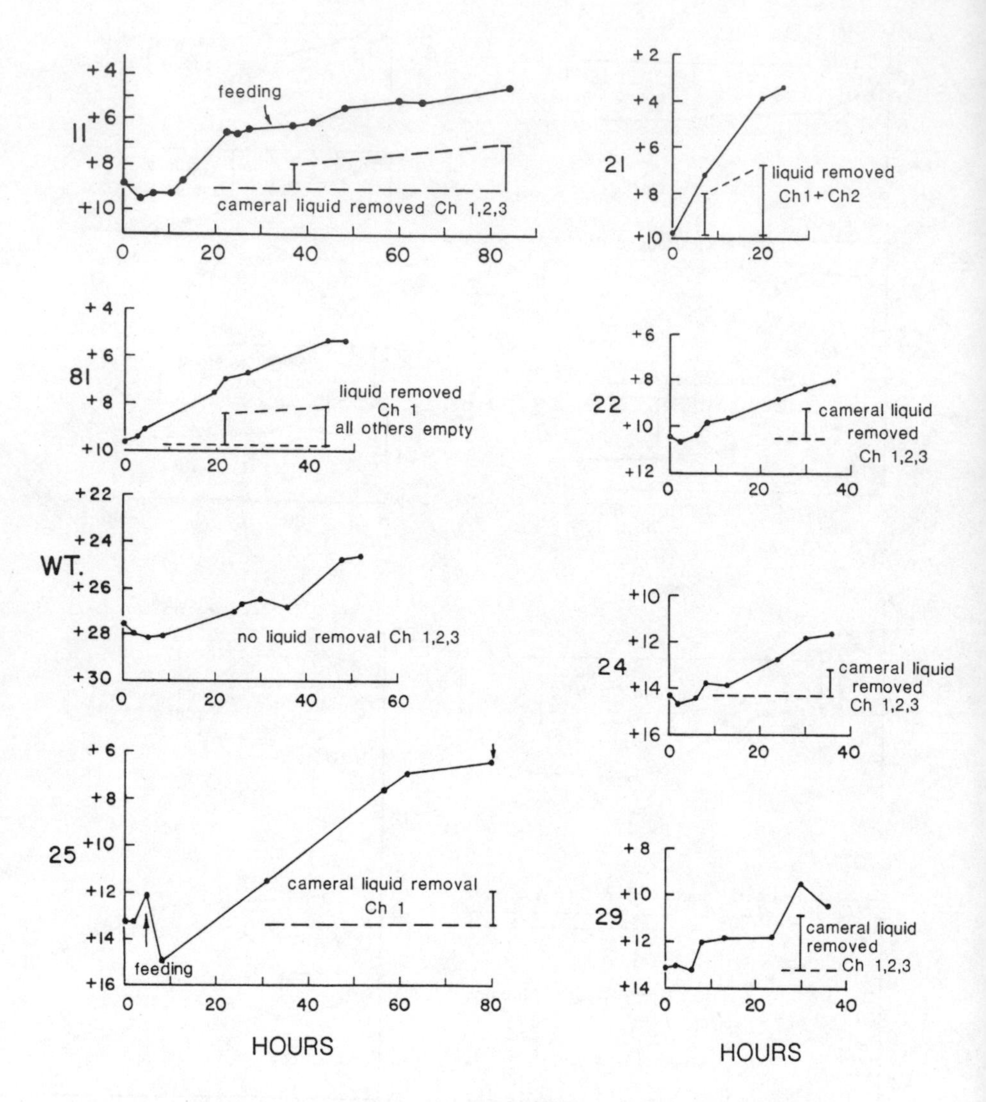

Patterns of compensatory buoyancy change in eight *Nautilus macromphalus* made artificially less buoyant by the addition of seawater into chambers or the addition of metal weights to the shell. The "liquid removed" part of the graph refers to the amount of compensatory buoyancy change that can be attributed to measureable cameral liquid removed.

wrapped, causing it to retract tightly into its shell, and was then transferred underwater into the box. This operation allowed me to detect differences in the animal's seawater weight repeatedly, to the nearest tenth of a gram.

The first step was to observe specimens whose buoyancy systems had not been disturbed. These specimens served as my controls. As expected, they all weighed about 1 to 4 grams in seawater. Over time I found that these weights varied slightly as the individuals ate or defecated; however, there were never large swings in buoyancy.

The next step was to produce a rapid change in the nautilus's buoyancy, then observe the results on the balance. The first of these experiments involved specimens that had suddenly been made heavy by the addition of liquid to their shells. By injecting 5 or 10 fluid ounces of liquid into various chambers, I could increase a nautilus's weight by 20 grams in a matter of seconds. As expected, the weight of these specimens gradually lessened as they pumped the liquid out of their chambers.

Far more interesting were experiments in which nautiluses were made suddenly *more* buoyant, which I achieved by breaking bits of shell material off the apertural region. These specimens were weighed periodically and yielded several fascinating and unexpected results. I found that nautilus responded to increased buoyancy by flooding previously empty chambers ten or 15 minutes after the buoyancy change was brought about. The flooding, however, occurred at a very slow rate, and the amount of change was ultimately quite limited; the animal adjusted its weight by a maximum of only 6 or 7 grams. Most of the flooding took place over the first ten hours of the experiment. After that, the rate of compensatory buoyancy change diminished and eventually stopped, even if the specimen was still quite buoyant.

These experiments on the effects of shell loss were central to understanding the life of this creature, because there is not a single nautilus shell that does not have a record of breakage. In some cases the size of a healed break indicates that the nautilus survived major damage. Most shell breaks are the result of predatory attacks by sharks, triggerfish, or groupers. Losing shell material makes the nautilus suddenly buoyant, and is thus the animal's Achilles' heel. If enough material is lost, the nautilus will float in uncontrolled ascent to the ocean's surface, where it will be at the mercy of various predators, lethally high water temperatures, and destructive reefs.

The buoyancy experiments conducted in 1982 showed that loss of as little as 5 grams of shell material would cause an adult *Nautilus macromphalus* to float to the surface of the aquarium if it ceased swimming. (If about 9 grams of shell material were lost, no amount of swimming could keep the animal submerged.) In these cases, however, the animal could resubmerge in 10 to 20 hours by refilling its chambers with liquid.

By incorporating radiographic observations with weight measurements, I was able to determine that compensatory buoyancy change occurs because of the addition of new liquid into previously emptied chambers. The investigation showed that liquid entered all chambers, not just the last two or three. This appears to be the reason a nautilus keeps its chambers unblocked and functional long after they have been emptied of liquid. A sudden need to increase seawater weight can be met much more rapidly if all chambers are operable.

Many hours were spent in front of Brainweigh, which faithfully yielded thousands of individual observations of nautilus seawater weight even though it sat over a rapidly evaporating saltwater aquarium in the tropical heat and ran on converted 60 cycles electricity from a 220-volt source. Brainweigh disappointed me only once, when a power surge during an electrical storm knocked out fuses in the balance with great sparks. Brainweigh made it back to the United States, but when it was finally unpacked and tested, it turned out to be stone dead.

The buoyancy experiments conducted over several years in New Caledonia yielded an interesting picture of the way nautilus uses its chambered shell. Throughout its life, nautilus strives to maintain steady, nonfluctuating buoyancy. During growth, liquid is removed to compensate for weight increases caused by shell and tissue development. After growth stops, ideally a nautilus would have to use its buoyancy system for nothing more than maintaining neutral buoyancy in the sea.

We also discovered that rates of liquid removal and addition are so slow that they are useless in powering vertical movement. A sudden shell loss can be compensated for, but once again, only at slow rates; the ultimate amount of buoyancy change possible in response to such damage is quite limited. Ironically, nautilus's compensatory buoyancy system is most efficient in shallow water, where it does not live and enters only under the cover of darkness, probably to avoid predation by fish. I believe that nautilus stays at the bottom fringe of its operable depth range because the risk from predators in shallower water is too high. And if nautilus is pushed much deeper, it will be gone forever.

Chapter 9

THE DEMONSTRATION OF VERTICAL MIGRATION BY NAUTILUS

No theory about nautilus has been quite as persistent as the one that claims these animals undertake large-scale depth migrations each night of as much as 1,200 to 1,300 feet in a matter of hours, and that they travel horizontally as much as a mile. Do such journeys really take place?

A vertical migration would begin in the great depths far below the sunny reef itself, on the vast, muddy plateaus of the deep fore-reef slope, where the brightest rays of the sun create at best a dusky twilight. As the afternoon advances even this faint light fades into complete blackness long before the sun sets. Nightfall changes this quiet place to a still darkness where thousands of dormant nautiluses, carefully concealed among the sparse reef rubble strewn amid the expanses of mud, slowly extend eyes and tentacles from their shells in preparation for the coming journey. Like hot air balloons, they lift off from their safe refuges in the covering blackness and begin a silent voyage toward the top of the reef.

The nautiluses climb by following the bottom contour, never moving more than a few inches above it. The gradient that leads from their daytime resting places at depths of about 1,200 to 1,400 feet to the upper parts of the reef is at first slight. In their first hour of travel the nautiluses cross sloping fields of mud pocked with the burrows of large shrimp. As the slope gradually steepens, the mud gives way to alternating fields of sand terraced by walls of reef talus and long-dead reef scarps. Soon the nautiluses must ascend nearly vertically, past and through the forests of gorgonian and alcyonarian corals that cling to the sides of the steep underwater cliffs.

Two hours into their long voyage the nautiluses reach the highest zones of the reef, moving past daytime predators such as triggerfish and groupers that have been rendered harmless by the night. When they finally reach the vast funnel-like openings of the reef's spur-and-groove systems that will lead them to the top of the reef itself, they begin to detect the faint odors of carrion not yet claimed by nocturnal sharks and crabs, or of their favorite food, the newly molted exoskeletons of spiny lobsters. Following these scents, the nautiluses converge on their prizes, settling like vultures onto the crustacean remains. Using their massive, parrotlike jaws, they feed on the molts, sending cracking sounds far into the underwater night as they fragment and ingest the calcareous exoskeletons of the lobsters.

As the night wears on, the nautiluses continue to forage both carrion and live prey, including hermit crabs, which they pull from their shells. As morning approaches, the sated nautiluses move back over the reef edge and fall freely down the sheer walls to their home on the muddy plains. When daylight again reaches the deep, the nautiluses are safely buried in their shells, silent and waiting.

Several men have lived this voyage. Arthur Willey, the great English zoologist, certainly followed it in his imagination, perhaps while resting in an outrigger on lush, tropical nights in Milne Bay, or while listening to Lifou's howling trade winds. How Willey came to his startling conclusion about vertical migration we will never know, for he left no clue in his journals or publications. In several memoirs he speaks without hesitation of vertical migration by nautilus, an observation whose effect on the interpretation of the many other fossil chambered cephalopods is inestimable. Right or wrong, Willey's theory completely changed our conception about the mode of life of both nautilus and its extinct relatives. The concept of vertical migration, put before the world without a shred of supporting documentation, was universally accepted on good faith by a normally surly and doubting scientific establishment. Every discussion of nautilus itself, and most of those dealing with the lives of the extinct ammonites and nautiloids, includes a reference to vertical migration.

I too have followed the nightly voyages of these creatures, first as a diver, watching the silent white disks glide into my lights, and later, huddled in the bottom of a boat with hydrophones dangling into the sea, listening to the faint ticking of an ultrasonic transmitter firmly mounted on the shell of an ascending nautilus. In three different countries I have listened to these man-made signals rising from the deep, sometimes over the gentle lap of equatorial waves, and sometimes against the howling nighttime wind in the trade latitudes. Willey was correct. There is now enough evidence to confirm that at least two different populations of mature nautiluses make fairly regular dawn and dusk voyages involving great depth changes. I wish there were some way to know how Willey came to his remarkable conclusion, one that could be substantiated only by teams of men armed with computer-age technology.

Vertical migration by nautilus has been confirmed by two sources. First, divers making nighttime excursions on the outer reefs of New Caledonia and the Loyalty Islands have observed what native fishermen have long known: Soon after nightfall large specimens of *Nautilus macromphalus* can be seen in the shallow areas of the outer barrier reef. The animals never venture into the lagoon, and have not been seen in similar areas during the day. Although it is possible that they spend the day hiding in the cracks and crevices of the upper reef, I think this is unlikely, or a rare occurrence at best. Moreover, large double scuba tanks have allowed divers to move down along the reef walls at night to depths of 100 to 200 feet, where they can meet the ascending nautiluses. It was during such dives that the creature's astounding behavior toward lobster molts was observed.

The most common spiny lobsters in New Caledonia and the Loyalty Islands belong to the genus *Panulirus;* individuals sometimes weigh as much as 10 to 20 pounds. These lobsters generally live in the top 30 to 40 feet of the outer reef. At night they come out of their secure lairs within the reef itself to feed. Although the fresh molts, or exuvia, of these large crustaceans would hardly seem a favored food, they are what the nautiluses seek in their diet, not the lobsters themselves. Captive nautiluses in an aquarium environment cannot catch and kill spiny lobsters. They will, however, readily consume all but the hardest parts of lobster molts, and there is now good evidence that suggests at least some of the calcium used in nautilus shell formation comes from such food sources.

Although the numerous observations of one species of nautilus in shallow waters of Melanesia suggested that vertical migration

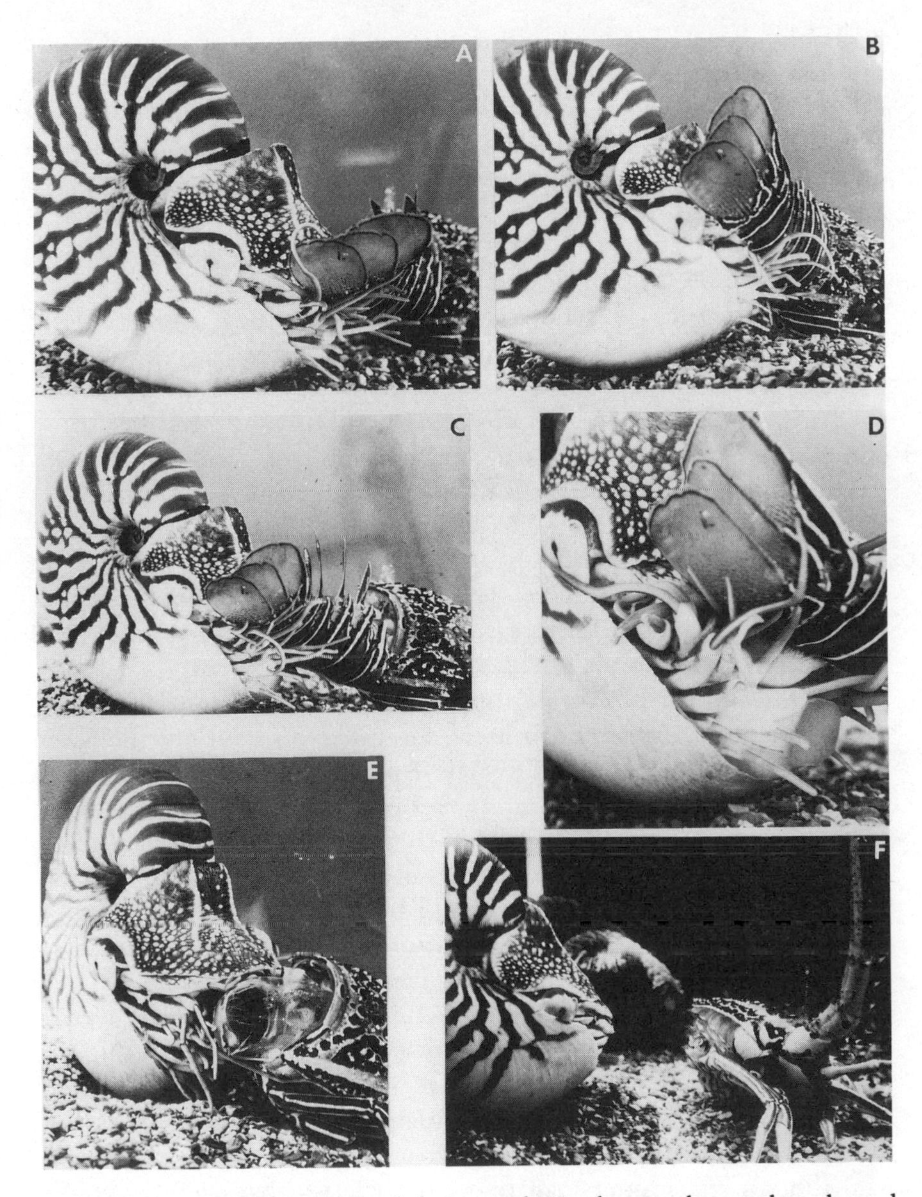

Nautilus macromphalus eating a lobster molt. In the two hours that elapsed between photo A and photo F, the nautilus steadily consumed the entire abdominal region of the lobster molt and gained 6 grams of in-water weight as a consequence.

probably occurred at least within this geographic area, they did not provide sufficient evidence to either confirm or invalidate Willey's hypothesis. Nautiluses could be found underwater, but the divers' bright lights precluded any sort of normal behavior following discovery. To substantiate the vertical migration hypothesis, we needed some means of remote telemetry so we could actually follow the movements of the nautiluses without disturbing them.

This required sophisticated technology. A transmitter had to be developed that could send a continuous positional signal strong enough to be received by hydrophones in a boat overhead. The transmitter had to be small enough to be mounted on a nautilus shell and still allow normal swimming and buoyancy, but be strong enough to withstand the great ambient pressures of the animal's daytime habitat. Because the transmitter had to be operational for at least several days, it required batteries of small size but great power. Finally, the observers in the boat had to be able to stay in position over the moving nautiluses day and night and in a variety of weather conditions, thus necessitating a weatherproof receiver and logistical support for the trackers.

In the late 1970s several groups of scientists began investigating the problem of using transmitters mounted on nautilus shells. In 1977 a transmitter developed by the Applied Physics Laboratory of the University of Washington was mounted on the shell of a single *N. pompilius* captured in Fiji. This transmitter gave only position; depth had to be inferred, and then only if the boat was right over the specimen so that a reading could be taken with an echofinder. Unfortunately the transmitter's batteries died the first night of tracking and no reliable information was obtained.

A better transmitter system was developed at about the same time by Jim McKibben, a technician working for Dr. Don Nelson of California State University, Long Beach. Nelson's team tagged and followed sharks in Tahiti, and many of the methods later used in nautilus tracking were derived from these earlier experiments. McKibben's transmitters ran on two batteries that were powerful enough to give them a life of up to two weeks and a range of up to two miles. Their greatest advantage was that they gave continuous, direct depth information as well as geographic position. A strain gauge to measure pressure was inserted in the transmitter. Increasing pressure, caused by increasing depth, lengthened the time between ultrasonic pulses given off by the transmitter. By calibrating the transmitter beforehand at known depths, observers could then calculate depth by the number of pulses received over a given interval of time. The performance of the transmitters varied widely, but they were capable of discriminating depths of as little as 50 feet. Detecting a 50-foot depth change by an animal that is more than 1,500 feet beneath the surface is extraordinary accuracy.

In 1982 the McKibben transmitters were used successfully for nautilus for the first time by a team led by Bruce Carlson of the Waikiki Aquarium. On the morning of June 19, 1982, they captured two large nautiluses in Palau and fitted them with transmitters. One transmitter failed almost immediately, but the other remained operational. The tagged nautilus, released by divers, swam down the steep Palauan reef wall and disappeared into the blue water at depths impossible for divers on compressed air. On the surface Carlson and his crew followed the animal's descent on hydrophones. Diving ever deeper, escaping the dangerous daylight, the nautilus descended more than 1,000 feet before it found refuge. Carlson, McKibben, and fellow diver and photographer Mike deGruy settled down in their small runabout boat and waited for nightfall.

At about five in the afternoon the by now thoroughly bored trackers began to detect movement by the nautilus. As the tropical afternoon changed to dusk, the observer at the hydrophone detected an increase in the number of ticks per minute being transmitted by the nautilus far below. The tagged specimen was ascending—and rapidly. Between six o'clock, when the equatorial sun was plummeting toward the horizon, and seven o'clock, when there was not a trace of light in the sky, the nautilus rose over 300 feet. The trackers now faced new problems. The nautilus, now foraging over the reef areas, was able to swim quickly out of the receiver's range, and the trackers had to search for the animal while trying at the same time to avoid the shallow reef. In the calm night the reef, so visible during the day, completely disappeared. Tracking continued, however, and during this first night the weary scientists were impressed by the amazing speed and distances their nocturnal quarry traversed.

As this remarkable experiment progressed into its second day, new elements in the drama were coming into play. To the north of Palau, near the Philippine Islands, a large atmospheric gyre system was coalescing into a low-pressure cyclonic disturbance. Completely atypical of this area and time of year, the tropical depression strengthened as it hurtled toward Palau. By the morning of the second day of nautilus tracking, tropical storm Ruby, as it was now classified and named, approached the islands. The nautilus, secure and calm in its deep daytime resting place, was unaware of the havoc overhead. In the face of hurricane winds and violent seas, the trackers were forced to flee for their lives.

For four days the winds raged in one of the largest disturbances ever recorded in this area of Micronesia. During this time the trackers occasionally sortied out from the secure anchorage of the Marine Mariculture Demonstration Center, their base of operations, to the tracking area, some seven miles away. They were able to locate the tagged nautilus but never could stay over the animal and track it continuously. The storm faded by June 25 and the team resumed its work, but heavy swells still prevented the researchers from tracking the animal for one solid 24-hour period. Nevertheless, the data collected showed that the tagged nautilus's average depth during the day was greater than its average nighttime depth.

The work of Carlson and his colleagues suggested that at least this species of nautilus was capable of vertical migration. Before any generalizations about this kind of movement could be made, however, the results needed to be confirmed by continuous tracking of many more specimens. Unfortunately Carlson and his crew were out of transmitters, time, and money. They left Palau with their tantalizing but incomplete data.

PALAU, 1983

A year later, on the afternoon of June 16, 1983, Carlson, Michael Weekley, and I boarded an overloaded Air Micronesia 727 and left Honolulu Airport for Palau with our 18 pieces of excess baggage. On this expedition we hoped to answer fundamental questions about vertical migration by nautilus. Does vertical migration take place in all nautilus species, and if so, how frequently? Do the animals move only during the night, or do they move during the day as well? What are the magnitudes of vertical depth change? Can any reason for vertical depth changes, if they occur, be deduced from physiological or ecological information?

Our Palau expedition in 1983 benefited greatly from Bruce Carlson's previous experience. One of the major problems he had encountered during his previous attempt at nautilus tracking was navigating at night. Because the tagged nautilus specimen traveled up the reef wall at night, the tracking boat had to move close to the reef to stay in contact with it, inviting potential disaster. To avoid this possibility we decided to mark the reef with waterproof strobe lights. Late each afternoon we would change the batteries of these strobes, which were tethered on buoys in the shallow reef. The twinkling lights, spread out along three miles, brought the reef to life each night, creating the impression of a metropolitan airport. Palauan fishermen must have looked out on the spectacle of this long flashing arcade in wonder. Our American technology was evidently coveted; at night we would hear the gentle throb of outboard motors as one by one our strobes were stolen, and the reef returned to its original, treacherous darkness.

The crew on the first tracking expedition had also suffered from fatigue. The boat they had used was an open, 18-foot fiberglass vessel that resembled a Boston whaler. Although fast and seaworthy, it was not designed for extended stays in open ocean. It had no provisions for cooking or sleeping, and offered no protection from the rain squalls. For the 1983 expedition I hired two boats, one to track, the other, a much larger craft, to serve as shelter during the night.

As my stay in Palau lengthened, my respect for Bruce Saunders grew. I began to understand the difficulties of keeping electronic and marine equipment operational in both the physical and political climate of Palau. By the time of our 1983 visit, the Palauan marine lab had been partly commandeered by Palauans in search of free rent. One of the dormitories had become an apartment building for locals, and all the air conditioners used in the laboratories had been appropriated for personal comfort. We arrived expecting that our boats would be ready; they were there, all right, but they were far from operational. The smaller, open boat had a 50-horsepower outboard motor, and the larger boat had a small diesel engine. We found that the electric starter on our outboard was not working. "The parts are on order!" we were told. Impatient to begin our study we set out anyway, with Mike Weekley, the weight lifter, starting the large motor each time with a rope.

We set two traps outside the reef, returned to land, and prepared our provisions for an extended stay at sea. The next day we returned to sea in our two boats and hauled in our traps. I was amazed by the size of the animals we caught; nearly a foot in diameter, Palauan nautiluses are among the largest in the world. We attached transmitters to two specimens and began our experiments.

In Palau the water temperature at the surface is in the high eighties and lethal to the nautiluses. To ensure that the tagged specimens were returned to cooler, deeper water as quickly as possible, Carlson and I carried them by hand over the edge of the reef, diving down through the clear water. The Palauan reef, one of the steepest in the world, is a vertical cliff that descends so sharply that water depths of 1,500 feet can be found only several hundred feet from its edge.

As Carlson and I dropped down the water temperature quickly became cooler. In the stillness I felt increasing euphoria as nitrogen began to work on my brain. We released the nautiluses when we reached 150 feet, and hung there as they began a lazy descent. As our specimens swam slowly downward, a large, predatory trigger-

The author tracking nautilus, Palau, 1983.

fish appeared from the reef edge and swam toward one of them. Carlson swam quickly downward, repelling the triggerfish. I was more concerned about a black-tipped reef shark that was following the proceedings with apparent interest. The two nautiluses descended further, dwindling white orbs that were soon in depths we could never visit.

Carlson and I rose back into the light and warmth above. In the boat Mike Weekley was excitedly listening with the hydrophone to the downward progress of our tagged specimens. The transmitters on these two specimens had slightly different frequencies so that each could be distinguished. For an hour they dropped, and as they eventually encountered a less steep gradient, they began to move off the reef and away from us. They stopped at a depth of over 1,200 feet.

The excitement of the initial capture of these specimens, followed by their tagging and release, had brought us to late afternoon. Our larger boat was anchored to the reef edge and the smaller boat tied alongside. To make continued readings, we had to take the smaller boat some distance off the reef. As the sun began to set and darkness gathered, we noted activity from the two nautiluses, which had lain motionless for several hours. Both signals began to change pattern. Within a half hour there was no doubt that the nautiluses were ascending, and rapidly.

By the time night had fallen, the two nautiluses had risen from a depth of 1,000 feet to about 500 feet. We took turns in the smaller boat on the five-minute ride to where the nautiluses could be detected. As night wore on, however, we began to have problems. We had released our two nautiluses at one end of a large bay formed by the reef. Because the reef in this area was oriented in such a way that it sheltered us from the prevailing winds, the sea near where we were anchored was usually much calmer than it was around the nearby point. Unfortunately, our nautiluses had begun to swim with considerable speed toward this point. Each trip we made to position ourselves above the nautiluses for an additional reading took progressively longer, and soon our transmitter-rigged specimens were two miles down the reef from their release point. Amazed by their swimming power, we began making nervous jokes about the large shark we were probably tracking with two nautiluses in its stomach.

Our larger boat, securely anchored to avoid piling onto the

nearby reef in a sudden squall, turned out to be an even lower-class hotel than we had imagined. For protection the living area was covered with an old tarp that, in spite of our best attempts at patching, was not waterproof. The boat was in the charge of a large Palauan named Marcus, an employee of the Palauan marine lab who had graciously consented to stay at sea with us. (It turned out later that he was charging double pay for this service, which was eventually billed to me.)

By midnight we were all thoroughly sick of nautilus tracking, and the initial excitement was replaced by a dull fatigue. Carlson and Weekley were sleeping under the tarp, so I decided to make the midnight reading alone. The ride through the tropical night to the spot where the nautiluses had last been tracked the hour before was exhilarating. Here at last was true adventure, I told myself, as I hurtled at high speed toward the point of the reef. I stopped the boat several miles down the reef from our anchored vessel, turned off the engine, and rigged the hydrophone.

The Dukane hydrophones we used consisted of a pair of headphones connected to a directional microphone that was put into the sea. The signals it transmitted consisted of sharp clicks whose time varied with depth. When the signal of either of our two nautiluses was detected, we used a stopwatch to determine the amount of time between the clicks. This operation was repeated ten times to arrive at a mean figure.

The hydrophone picked up far more than just the nautiluses. The sea was alive with sound. I could hear the clicking of dolphins and fish and the cracking and popping noises of snapping shrimp. The daytime ocean was relatively quiet, but at night the sounds intensified. The snapping shrimp were the worst then. They were so loud that if one of the nautiluses moved among them, the signals it transmitted could not be heard.

On this first night I wanted to make my reading and hurry back to the larger boat. The one nautilus had taken me well past the part of the reef that had screened us from the wind. The wind had risen, and even in the ten minutes it took to find and measure the depth of our nautilus, my boat had drifted out to sea quite a bit. I pulled the starting rope on the massive engine and nearly fell out of the boat: The rope had broken. This was trouble. I spliced the frayed rope back together, but it broke again. I searched for whatever line was in the boat—the anchor rope, shoelaces, cord from my raincoat. Each broke in turn. I was steadily being blown out to sea, the next stop being the Philippines. In between splice attempts I flashed my light toward the dim glow that marked the position of our anchored boat. In my cleverness in sneaking off and not waking the others, I had brought no water, food, or oars.

Two hours later I heard the dull throb of an engine. A flashing light approached. Our larger boat had sallied forth to find the smaller boat. Marcus assured me that he had never lost sight of my light. He didn't want to lift anchor unless it was totally necessary. Weekley and Carlson thought it was hilarious. Welcome to Palau.

Our life became monotonous and tiring. The incessant rain squalls had a knack of finding us during moments of sleep. The only dry spot in the larger boat was a small oily space next to the diesel engine, a place of honor reserved for Marcus. For hours at a time we would alternate turns on the hydrophone, taking a reading every 15 minutes, listening to the clicking of our specimens against the background noises of the sea. I think we all became more than a little crazy as sleeplessness, the constant squalls, and canned food made inroads into our sanity.

Meanwhile the nautiluses became clever adversaries, using acoustic tricks to elude our sonar search that would have impressed an enemy submarine captain. Sound in the sea is affected by many factors. On some nights or days the nautiluses were easy to track. Other times combinations of thermoclines, which can bend sound waves, and the covering noise of snapping shrimp or the crash of waves on the reef prevented us from keeping up with them. The nautiluses would silently slip into soundproof walls of the great reef system and then sneak along while we listened frantically through sounds of the sea for our quarry. Sometimes we lost specimens for days at a time and had to increase our search radius gradually until the skulking culprits gave themselves away.

After a prolonged period of loss, contact often came as the barest hint of the regular clicking, borne through miles of sea by a providential thermocline, or density layer. We would then put the volume at maximum to try to filter out the myriad noises picked up by the hydrophones or lent to us by yearning imagination, and strain to get a direction. When we established a direction, we would then race after it in our tracking boat. Such rapid flights, though joyous during the day, were filled with terror at night; who knew what logs or part of reef lay hidden in our path? These moments of discovery were the high points of days that were mostly filled with boredom.

Each animal was affixed with a transmitter of a different color. We lost Green the first night of tracking; the clever specimen hid in the reef at 400 feet surrounded by snapping shrimp that obscured his sonic signal, then sped down the deep reef wall for several miles. We found Green sometime later by accident, long after we had given him up. Blue was a voyager. We followed him for two weeks as each day he steadily moved several miles south along the reef toward the island of Peleliu. We had to venture farther and farther to find him. Red was a mystery and lasted only two nights, the shortest time. On a blustery night in heavy seas he moved up onto the rocky, shallow reef point, where background noise is the highest. His signal weakened, and we were sure he was eaten by a shark. Perhaps Red just eluded us, but we never found him again.

And finally there was Yellow, who confirmed our hypothesis. For seven days and nights Yellow was never able to shake us. He rose each dusk and descended each dawn with such regularity that we joked about setting our watches by him. Yellow spent every day at depths of 1,000 to 1,200 feet, resting or slowly moving on the vast, muddy plains a mile offshore of the steep, rocky Palauan reef walls. Each afternoon between four and five o'clock he began his journey inshore toward the walls of the reef. In the equatorial latitudes of Palau the sun hits the horizon at 6:15 P.M., and not long after, Yellow was on the reef at depths between 200 and 300 feet. He was ruled by the night; with the first glimmer of daylight he would cease his wandering along the reef walls and begin to swim offshore toward deeper water. By the time the sun heralded a new day, Yellow was again on the deep muddy bottoms, quiescent, dormant.

One night Yellow was fooled by the full moon, and as a result taught us a great deal about the stimuli that provoke vertical migration. At about nine o'clock on the third night we were following Yellow, the nearly constant squalls that had been soaking our open boat since late afternoon began to disperse. The sky, which had been completely overcast, opened and lightened and finally revealed a glorious full moon. With the last departing squall, Weekley and I saw our first wonder of the night—a moonbow, dim but distinct, attesting to the brightness of the tropical moon. Yellow was not amused. Over the next hour we listened to the lengthening signals of a panicky crash dive as Yellow began to swim deeper in the face of this all too short night. He dived several hundred feet before he stopped to figure this out. By midnight he was back into shallower water but was deeper than usual, and in our minds, cranky as hell at the vagaries of an unpredictable universe.

After observing seven days and nights of Yellow's vertical migration, as regular as Willey had ever envisioned, we returned to land on shaky legs for hot food, showers, clean clothes, and sleep. During our first night ashore the winds rose in a sudden, larger than normal squall. We were chased from bed by the panicky director and discovered that our anchored boat was swamped and sunk, most of our equipment gone, and the motor drowned. Two days

Mike Weekley (right) and the author after seven days and nights tracking Yellow in Palau, 1983.

later we got the salvaged boat running again and set out after Yellow. The sea was devoid of his clicking. We never found him.

The four specimens tagged in Palau followed a fairly regular habit of twice-a-day depth change. There were certainly irregularities, but in general the movement followed an offshore-onshore pattern. The vertical movement of the nautiluses occurred as the specimens followed the bottom contours in this twice-a-day, offshore-onshore migration.

Although the most fundamental contribution of these experiments was the demonstration of regular nightly migrations, we also derived useful information from average habitat depths. We calculated that nautiluses would ascend to as little as about 200 feet and dive to a maximum of 1,500 feet, but they would not stay at either of those depths for any great length of time. By integrating the long-term depth curves, we found that the *average* habitat depth over any 24-hour period was between 700 and 900 feet. As we had previously found in the chamber-liquid-emptying experiments, these are the maximum *average* depths nautilus can inhabit and still keep its shell empty of liquid.

The transmitter studies revealed a great deal about the rate of ascent and descent, as well as about rates of horizontal swimming speed. During the two weeks we tracked Blue, we found that he moved along the reef in a southwestward direction and traveled about 20 miles along the coast. On our last day in Palau, Blue was still sending information as he moved on and offshore.

Our transmitter studies in Palau demonstrated that at least one species of nautilus showed regular vertical migration. However, many questions remained. The most pressing concerned the habits of other nautiluses in other localities. Was vertical migration typical of all nautilus species?

NEW GUINEA, JUNE 1984

In some countries airports and airplanes have become mundane, accepted things, but in other countries, the arrival of an airplane is a great event, even if it is a daily occurence. Papua New Guinea belongs in the latter category. Many hours and stops after my departure from Fiji, I wearily stepped off my Air Pacific flight to face the

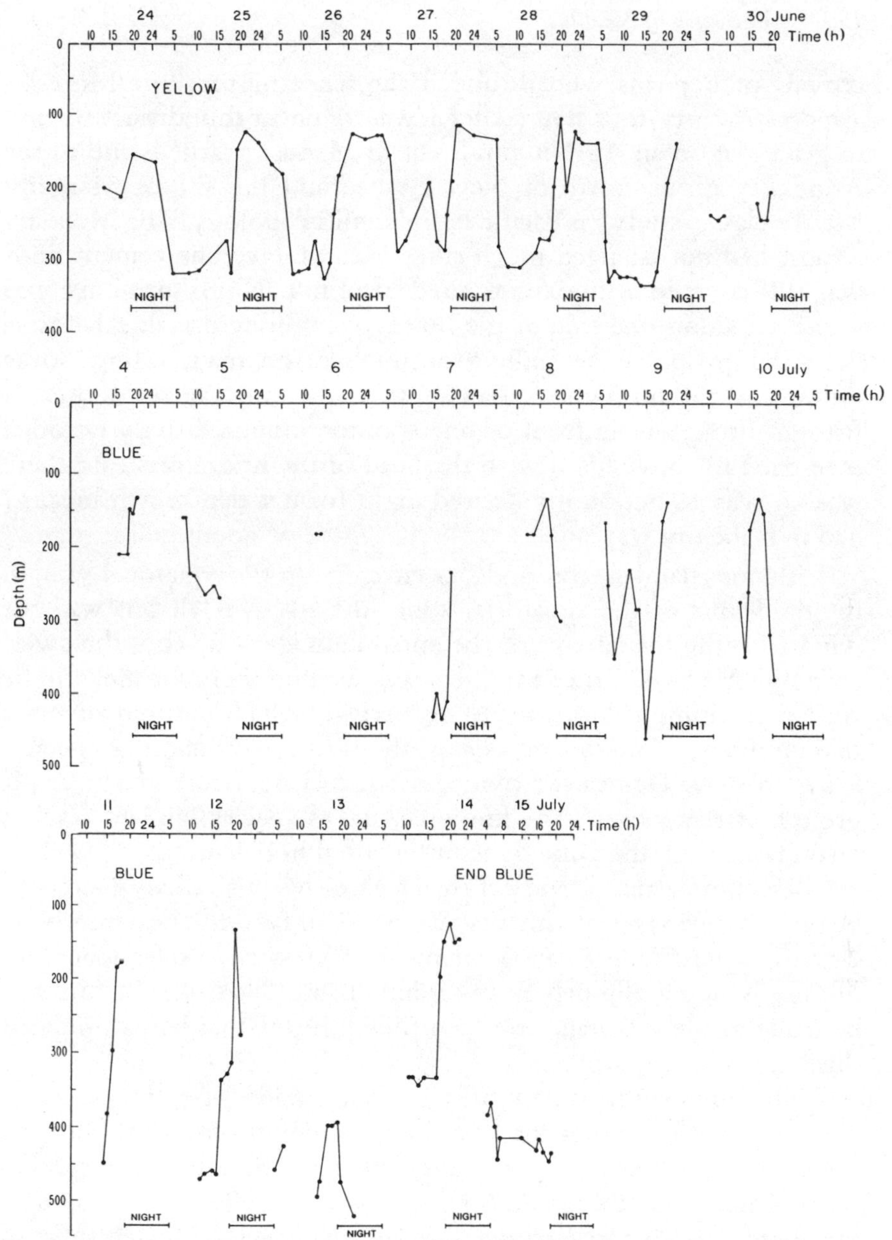

Vertical movement with time of two *Nautilus belauensis,* Yellow and Blue. The ordinate records the depth, as derived from the strain-gauge depth sensor of the transmitter. Yellow was tracked from June 24 to June 30, 1983. Blue was tracked between July 4 and July 14, 1983. Each dot represents the mean 6–10 depth readings taken over a 2–3 minute period. Numbers on the horizontal axes refer to the hour of day. Periods of darkness (night) are indicated by bars at the bottom of each graph.

crowds of Papuans who thronged the fence of the Port Moresby airport. Twenty-four hours later I was again in this airport, trying to get a seat on an Air Niugini flight to Manus, a large island of the Admiralty group north of New Guinea and the site of Margaret Mead's classic study on Pacific Islander anthropology. The Moresby airport had not changed much since Mead's day. The cement floor was still covered with expectorated betel nut. Heavy drinking was going on along one side of the huge, open-air terminal. At least a thousand people were milling around in the main lobby, some fighting to get on flights, others apparently enjoying the sights. A fistfight broke out in front of one counter when a heavily tattooed man tried to shove his way to the head of the line. I was told there was no way to get on my desired flight for a week, which meant I had to bribe my way on.

Manus Island is large and sits virtually on the equator. Lying in the doldrums of the equatorial zone, the island is all but without wind to ruffle the surface of the surrounding sea or cool the sweltering jungle heat. Bruce Saunders was waiting there for me, and he waited in triumph. Just prior to my arrival he had captured alive the first complete *Nautilus scrobiculatus,* the rarest of all nautilus species. I was in New Guinea for two reasons: to help Bruce continue his project of studying all the known nautilus species, and to track a second one with the same transmitters used in Palau.

We chose a tiny island off the coast of Manus called Ndrova to begin our tracking experiments. We were, in fact, not far from New Britain, where Willey had spent most of his time. Perhaps because of this, Willey was much on my mind during this trip. The findings he made in New Britain provided the incentive that had stimulated this latest voyage of mine.

The quiet, calm waters off Manus were ideal for trapping and studying nautiluses. For the first time, we could track without constantly battling the wind and waves. We soon found, however, that there were worse things than wind and waves. The heat was unimaginable and we had to drink large quantities of water to keep from dehydrating. We were all taking massive doses of malaria tablets—chloroquine and Fansidar, the latter turning us slightly yellow.

• • •

The health of the people in this region had changed very little from Willey's descriptions. In spite of the tremendous medical advances since Willey's time, many of the people of Manus still had appalling open sores on their arms and legs. Malaria was endemic. Dengue fever and dysentery were prevalent. Soon after my arrival I contracted a fever of some sort that made my entire stay unpleasantly dreamlike.

We traveled to Ndrova, our trapping site, in a small fishing boat lent to us by the very efficient New Guinea Department of Fisheries. Saunders and his field assistant had already trapped in this locality and knew that two species of nautilus could be captured there. Although the nautiluses in this region are large, they are not quite as large as those in Palau. Still, I had no problem attaching a transmitter to one of the newly captured specimens.

Surface water temperatures were even higher here than in Palau. The water was blood-warm and oily-smooth. Perhaps this is the reason the sharks in the area are so aggressive. The locals would not swim out into the water on the surface. Several times in the preceding week, sharks had chased Saunders and his field assistant out of the water. In a fuzzy, feverish state I readied my nautilus and prepared to dive with it, down out of the warm-water zone as we had in Palau. My colleagues refused to dive with me. One of the Manus islanders swore that he had finished a diving course and volunteered to ride shotgun. I gave him a bang stick and we descended into the cool of the sea.

At 100 feet my companion's eyes became as big as plates, and he hastily returned to the surface. I discovered later that prior to our dive he had never been deeper than 20 feet. I continued down to 150 feet, where I dropped the nautilus and then watched for a while as it made its way down into the depths. I returned to the surface without seeing a shark.

The tagged nautilus took off like a shot. It began swimming along the coast of the small island, descending all the time. Ndrova is separated from Manus Island by a strait of water too deep for nautilus to traverse, so the animal could only circle the island. Within an hour it had traveled a mile; then it dived over a steep reef face into very deep water.

During the long afternoons I used the Fisheries' vessel, which fortunately did have cover from the tropical sun. As evening came, however, I changed to another boat for night tracking. The Ndrova islanders still built dugout canoes, although modern times had brought them outboard motors for their bigger craft. The local chief had volunteered his large outrigger for the nighttime trackings, and most of the island's small population wanted to go to sea that night to see what this curious white man and his funny machines were up to. After listening to the nautilus ascend in the early evening in a manner similar to that of the nautilus in Palau, I transferred to a large dugout canoe to begin my long, somewhat fever-stricken vigil.

We rowed out over the reef on a clear, calm night in the company of millions of stars. The old man watched curiously as I assembled my hydrophone, with its large parabolic receiver, and started listening to the sharp ticking coming from the depths. Everyone on the boat listened in turn, patiently asking questions in

Bruce Saunders (right) with the author in New Guinea, 1984.

their perfect English about my thoughts and hopes for this project. Arthur Willey was on that dugout once as well, listening and smiling a knowing smile, or at least so I felt.

At dawn, after a night of foraging, the nautilus saw the light or registered whatever signaled the coming day, then crash-dived. He spent the day on the deep bottom and rose once more the next evening. But by then I was exhausted. That same evening Bruce Saunders decided to return to Manus, and the experiment was terminated.

But the experiment was a success. Because of the weeks of work in Palau, I no longer needed to follow any animal for many days. The tempo of vertical migration had been established in Palau. I now needed only to establish how common this behavior was among the widely scattered tribes of nautilus. In this second population the same behavior had been demonstrated.

Chapter 10
HOW MANY SPECIES?

As European civilization expanded, its philosophers and early naturalists began to become aware of the staggering diversity of the earth's animals and plants. The diversity was so overwhelming that the systems for naming these creatures started to break down, and eventually the same common names were used for many different species. This problem occurred in the case of nautilus, a name the Europeans used for two different forms. The paper nautilus, or argonaut, is a small octopus that lives in the Mediterranean and produces its shell in curious fashion, using greatly expanded tentacles instead of body tissue to calcify it. Because these large tentacles looked like sails to the ancients, they mistakenly thought the argonaut moved across the top of the water. The pearly, or chambered, nautilus was often confused with this creature. Carolus Linnaeus, the father of modern taxonomy, untangled this confusion by giving each animal a separate name. In 1758 he designated a new genus he called *Nautilus* and named *Nautilus pompilius* as its type species. He assigned the paper nautilus to the genus he named *Argonauta*.

In the eighteenth century European naturalists became aware of nautilus shells that were different from those of *N. pompilius*. One of these shells was orange and white in color, and quite dissimilar in shape. These discoveries proved that more than a single nautilus species existed. No one knew what other oddities related to this creature would eventually turn up in the poorly explored Pacific region.

There was also great confusion about the relationship between the modern shells placed in the genus *Nautilus* and the thousands of fossil forms that to some degree resembled them. It was obviously by their shape that many of these shells were kin to nautilus; some bore a strong resemblance. The common practice of eighteenth- and nineteenth-century paleontologists was to name any nautiloid fossil "nautilus" that even remotely resembled the modern shell.

Two questions emerged more than two centuries ago: How many species of *Nautilus* were there, and what was the geological age of the genus? The great English malacologist G. B. Sowerby addressed both questions and answered them with remarkable accuracy in his monumental *Thesaurus conchyliorum,* published in 1848. Sowerby succinctly commented on the various taxonomic problems of *Nautilus:*

> By the common consent of the more modern naturalists, this genus retains the name of *Nautilus,* although it is not very nearly related to the Nautilus of Pliny, which is called *Argonauta* by the moderns, to which the commonly received and elegantly embodied fiction
>
> "Learn of the little Nautilus to sail"
>
> of right belongs, and has no reference whatever to the present genus. Few are the recent species of *Nautilus* as yet known, five being the utmost, of which one may be only a sexual variation of the common *N. pompilius,* and one other has only very lately been discovered. Of fossil species there are many, belonging to most of the geological series, from the Carboniferous limestone upwards. We believe that all the fossil species are distinct from the recent. . . .

In this brief passage Sowerby outlined the understanding of the genus *Nautilus* that has remained to this day: The species are few in number (four or five), and cannot be identified with any of the fossil nautiloid species currently defined.

IS NAUTILUS A LIVING FOSSIL?

The question of whether or not nautilus can be considered a "living fossil" was a perplexing one. Scientists had long recognized that some animals or plants show rapid evolutionary change, whereas the fossil record of others seems to show no apparent change over long periods of time. Charles Darwin was beset by this enigma. According to his theory of natural selection, all animal and plant species should show gradual evolutionary change as less fit types are weeded out. Darwin realized that over long periods gradual shifts in environmental conditions brought about by geological and climactic change were the major agents necessitating morphological changes in most species. Why, then, did some species seem to show almost no change, even over very great stretches of time? If evolution worked as Darwin hypothesized, through the slow, gradual changes occurring within lineages over the millions of years of a species's history, no species with a significant geological range should be exempt. But clearly some were.

During the last century European paleontologists became familiar with fossil nautiloids that appeared to be similar to nautilus, especially those from the Triassic and later periods. Because of the morphological similarities between the fossil and modern shells, the name nautilus was given to all of these Mesozoic and younger fossils. Nautilus thus appeared to have a geological age of over 200 million years, making it, like the crocodile and horseshoe crab, one of the most ancient organisms still known to be living—and thus an archetype of a living fossil. There was only one nagging worry: Fossil nautiloids seemingly disappeared during the Miocene, a geological unit that ended about five million years ago. Rocks from the Pliocene and Pleistocene epochs were devoid of fossil nautiloids. This five-million-year gap was even more puzzling because of the tremendous number of Pliocene- and Pleistocene-age rocks lying around. Indeed, there is little chance of fossilized nautilus from the Miocene going undiscovered, because as we approach the present day, both the volume and exposure of sedimentary strata increase; moreover, these rocks contain significant quantities of oil and other economically important mineral resources, and are thus thoroughly searched. Where, then, were the nautilus fossils?

Paleontologists began to look at fossils with increasing scrutiny. "Similarity" was no longer a sufficient criterion for placing a newly discovered fossil within an existing taxon. As twentieth-century specialists reexamined the criteria used by their nineteenth-century scientific forefathers, they began to widen and increase taxonomic differentiation. As small but sure morphological distinctions became apparent, many fossil species put into the genus *Nautilus* were given new names. Two great cephalopod specialists, A. K. Miller and B. Kummel, concluded that only the living species could be called *Nautilus* and that all of the fossil species were surely related ancestors, but taxonomically differentiable from their present-day counterparts. Consequently, by the 1950s the hundreds of fossil species originally considered members of the genus *Nautilus* had all been transferred to other genera. Nautilus was certainly living, but it was no fossil. In fact, it had no fossil record at all.

HOW MANY SPECIES?

In his 1848 compilation Sowerby had recognized five species of *Nautilus*. Two of these, *Nautilus pompilius* and *N. macromphalus,* were identified by their soft parts as well as their shells. Sowerby thought one of his five species, *N. ambiguus,* was not a distinct species but rather a possible variant of *N. pompilius*. Most later researchers concurred with this opinion, and the name *ambiguus* has been dropped. Sowerby's other two species, *N. umbilicatus* and *N. stenomphalus,* were known from their shells only. No one had ever seen living specimens or had any information about the morphology of their soft parts.

As exploration of the Pacific Basin increased in the late 1800s and the first part of this century, thousands of new species of animals and plants were discovered and named, but the number of creatures identified as species of *Nautilus* changed little. Although several new names were proposed, based on shells from the Australia region, they were never accepted, and the valid number of species remained at four. The species name for one of these, *N. umbilicatus,* was changed to *Nautilus scrobiculatus* after it was found that "umbilicatus" had been incorrectly conferred.

Somehow the low number of *Nautilus* species led to the conclusion that the animals are rare. Thus, the idea of nautilus as living fossil was coupled with the notion that it was on the brink of extinction, a last relic waiting for final execution after a 500-million-year-long history. Even the fact that nautilus shells could be bought cheaply in any curio store did not dispel this notion. Investigators venturing out to the far Pacific to study nautilus were admonished by both their colleagues and the public about further endangering an already endangered creature. Clearly, more information was needed about the number of existing species and their population sizes. How many species really existed? How variable were the morphological attributes of these species? Was nautilus in final decline, remaining constant in number, or—heresy!—perhaps radiating anew and actively enlarging its distribution, populations, and number of species? These were the questions that prompted many of the expeditions during the 1970s and 1980s.

FIJI, 1975–1976

Near the end of my first research trip in 1975 to study nautilus, I met a fisheries expert from Fiji named Robert Stone. He was in Nouméa at the time to attend an interisland fishing conference. Stone told me that he was engaged in deep-sea shrimp fishing in Fiji and had a terrific problem. Nautiluses kept swimming into his shrimp traps and stealing his catch. Was I interested in observing this? I asked him how deep the traps were. About 1,200 to 1,800 feet deep, he replied. I was astounded. Although nautiluses were reportedly present in Fiji, nothing was known about them there. The deepest a live nautilus had ever been found was 1,800 feet, as recorded for a specimen caught by the nineteenth-century *Challenger* Expedition during deep-water dredging. Most specialists concluded that this record must have been an error on the part of the *Challenger* crew. But here was Robert Stone, who had evidence that nautilus could indeed be caught at great depths. I was very interested in seeing these deep-water nautiluses.

On my way home from New Caledonia I stopped over in Fiji for a week and witnessed the capture of several more specimens. I was struck by the small size of these animals, which looked like tiny *Nautilus pompilius*. At that time *N. pompilius* was known primarily from the Philippine Islands. Its average shell size is about ten inches in diameter, but the Fiji specimens were about five inches across. I caught a particularly small specimen, put it in a plastic bag with a battery-powered pump, and set out for North America.

My tiny nautilus immediately vomited all the fish bait it had eaten in the trap. I spent a long night in the Pan Am jet's kitchen straining its water through coffee filters and arrived in Honolulu very tired. Before changing planes, I had to pass Hawaiian agricultural inspection. Hawaii has necessarily tough laws restricting importation of many foreign plant and animal species. I was immediately seized when it was discovered that I didn't have a valid permit for my small companion. The precious nautilus was swimming happily about in its bag when the Hawaiian inspector confronted it. "A fish without a permit," he exclaimed. "This fish will have to be killed and you are about to get a big ticket."

"That's no fish," I explained. "Can't you see it's a snail?"

Trapping for nautilus, Fiji, 1975.

The inspector mulled this over. "No law against snails, but I've never seen a swimming snail. Passed."

Six hours later we were in North America. It was the first time in over six million years that a nautiloid had returned alive to that continent.

My nautilus from Fiji lived for a time in the Steinhart Aquar-

The author in Fiji with a specimen of the tiny race of *Nautilus pompilius*, 1976.

ium in San Francisco, and then died there. When I cut open its shell, I was astounded to discover that this small creature was not the immature specimen I had assumed it was, but a fully grown adult. Although my specimen resembled the familiar *N. pompilius* from the Philippine Islands, it was much smaller and had 27 chambers, not the 33 to 36 usually found in that species. Here was an indication that perhaps more diversity existed among nautiluses than was currently recognized.

In 1976 I returned to Fiji and spent two months trapping nautiluses from the fore-reef slopes near Suva. The nautiluses were plentiful, but could be caught only in depths far greater than those of the common trapping grounds off New Caledonia. Nautiluses were never found in shallow water at night, and Robert Stone and I never saw them on the several night dives we took. The nautilus could be trapped no shallower than about 400 feet; and the richest catches were from depths of over 1,200 feet. All of the mature specimens were small and showed little variation in anything but shell color patterns.

After this trip I studied the collections of nautilus shells in various museums around the world. I soon discovered that *Nautilus pompilius* could be found on many islands and continental edges spread across the western Pacific Ocean. The shells from each island group, however, had subtle but distinct differences, particularly in the diameter of mature specimens, but also in coloration, coiling, and ornamentation.

In 1977 Bruce Saunders and Claude Spinosa began their work in Palau, and in 1981 Saunders shocked a lot of people by describing the Palauan form as a distinct new species of *Nautilus*. He based his conclusion on the large diameter of the shell, its distinct chevron ornamentation, and other minor features. The long-held view that there are only a few, slightly different species of *Nautilus* began to unravel. In a few years two new, distinct morphs of what had previously been considered *N. pompilius* were discovered. In 1980 teams of Japanese investigators had begun intensively studying living nautiluses in several areas of the Philippine Islands and comparing them to the nautilus of Fiji. They added valuable new information not only about the Fijian forms, but also about the Philippine forms. The study of *Nautilus* species branched out to include investigations of protein and shell chemistry as well as classical morphological descriptions of shell and soft parts. However, several major questions remained. Two of the species known at least since Sowerby's time, *N. scrobiculatus* and *N. stenomphalus*, had been identified only from drifted shells. Were they truly distinct from the well-known species *N. pompilius* and *N. macromphalus?*

NEW GUINEA 1984

In early 1984 Bruce Saunders and I received a National Science Foundation grant to pursue the two elusive species, and to document the degree of variability within already known populations of nautilus. In addition to studying shell morphology, we were also collecting bits of tentacle to be examined by gel electrophoresis techniques in the laboratory of Dr. David Woodruff at the University of California, San Diego. This process yields estimates of genetic distance between species. The tentacle sampling was a way of getting tissue for our studies without killing the nautilus. We knew we could sample the familiar populations in the Philippines, Palau, New Caledonia, and Fiji. But where were the missing species?

The most unusual of all nautilus shells comes from *Nautilus scrobiculatus*. The only known observation of the soft parts of this rare and unusual creature had been made by Arthur Willey, who had been given a single specimen, already dead and partly decayed, during his visit to Milne Bay in New Guinea at the turn of the century. New Guinea seemed a reasonable place to begin our search.

Saunders and I considered trapping near Milne Bay, but soon received information that Manus Island, where we had conducted our transmitter experiments, might be a better place. Saunders had communicated with a Mr. Ron Knight, who ran a shipping company on Manus Island. Knight was a keen and expert shell collector and had noted that many drift shells of both *N. pompilius* and the rare *N. scrobiculatus* were to be found along the island's beaches. Saunders decided to begin operations there.

Saunders and his field assistant, Larry Davis, arrived at Manus Island in May 1984. They were immediately welcomed into Knight's house and family. Knight's son, Ronnie Jr., ran the local dive shop and also had access to boats and equipment. Aided by New Guinea Fisheries they began the nautilus fishing operations. Large specimens of *N. pompilius* were soon captured, but none of the rare *N. scrobiculatus*. They began to search farther afield.

Saunders decided that to find *N. scrobiculatus* he next had to try trapping at Ndrova, the site of an old German copra plantation, where many drift specimens of the rare species had been found on the beaches.

Local Manus islanders watched as Saunders began his strange form of fishing. Two white men with scuba gear dived into the sea with the end of a bright yellow rope, only to emerge several minutes later. Their boat slowly moved offshore into deep water, trailing rope out the back. The boat finally stopped some distance offshore, and a large wire and iron cage was thrown into the sea. As the boat turned toward Manus, the long, floating rope disappeared from the sea's surface as the trap quickly sank. The islanders paddled out to observe this strange behavior. No trace of rope or the trap was to be seen. They paddled back to where the white men had first dived into the sea, and in the calm water they could see the faint yellow of the rope below. Donning masks, they peered down. The rope was tied to a large coral head about 20 feet down and stretched out in a taut line offshore. It ran out beyond the wall of the reef and downward into the blackness of the deep water. The Ndrova fishermen quickly paddled back to tell their chief.

Two days later Saunders and Davis returned to pull up their trap. They dived to untie the line but were almost immediately surrounded by sharks. Back to back the two divers managed to get the line untied, and then they retreated into shallow water, where aggressive white pointers and black-tips circled them. Things were no better on the surface. Two stern Ndrova islanders told the nautilus fishermen that they were "requested" to meet with the Great Chief.

Papua New Guinea is an independent country with a large civil service and elected officials. But in many of the more remote areas, like Manus Island, the local tribal councils are still autonomous. Saunders had been given permission by both the Papua New Guinea government and the local elected officials to pursue his nautilus research, but all of a sudden he had a third government body that was definitely not pleased.

The Great Chief lived far down the coast of Manus Island in a village that was connected by a small dirt road to Lorengau, Saunders's base of operations and the largest town on the island. But the quickest way of getting to the village was by boat, and even this was a three-hour voyage. Saunders loaded his official papers into a dugout canoe and, accompanied by young Ronnie Knight, set out to meet the Chief.

The long, thin dugout, powered by a large outboard engine, skimmed along the coastline at high speed. Numerous seagoing crocodiles were startled from their shoreline perches as the boat moved south through intermittent rain squalls. Saunders finally arrived at the village and was taken in to see the Chief.

Bruce Saunders was raised in the American South and has the courtly manners of a Southern gentleman. He charmed the old warrior, assuring him that fish would not be scared away from Ndrova Island. The Chief was mollified and gave Saunders permission to continue his work. As Saunders got up to leave, the grizzled old man took him aside. "By the way, have you been having trouble with sharks?" asked the Chief. Saunders said they had been chased out of the water by sharks on their last visit to Ndrova. "I will attend to it," said the chief. "You will not be bothered any further by the sharks." Saunders, by his own admission a deeply superstitious man, left the audience in a subdued state.

Saunders and Davis returned to Ndrova Island to resume their trapping. Soon they were bringing in rich catches of *N. pompilius,* and just as the Chief had promised, they had no more trouble from the sharks.

In late May they pulled up a trap, and among the usual collection of nautiluses inside it was a large shell, deep orange in color. Saunders knew immediately that he had trapped the long-sought species. He was probably the first man ever to see a living *Nautilus scrobiculatus.* Great whoops of joy echoed out over the still water, much to the astonishment of the Ndrova islanders.

The first *Nautilus scrobiculatus* was an amazing sight. We knew in advance that the creature would be different in form, but we had no idea that the shell itself would bear an additional, totally unexpected, surprise. When Saunders reached into the nautilus trap to pull out his prize specimen, he found that the bottom of the shell was obscured by a thick layer of 1/4- to 1/2-inch-long fur. *Nautilus scrobiculatus* had saved a great surprise for us; it was covered with an outer shell layer—the periostracum—that was more extensive than and very different from any such feature in other known nautilus species. This layer had never been preserved on any of the drift shells.

By the time I arrived at Manus Island a week later, Saunders and Davis had captured several more *Nautilus scrobiculatus.* We captured three on my first day at sea and several others over the next week I was there. Ndrova Island was the first place on earth where more than one nautilus species was known to inhabit the same area.

Saunders told me the story of his meeting with the Great Chief, and how the Chief had commanded the sharks to stop bothering him. "Right," I replied. On the eve of my last day around Manus, Ron Knight's son and I traveled to Ndrova in one of the dugout canoes. Young Ronnie was a free spirit with a great dragon tattooed on his arm. Our goal was to photograph the beautiful reef walls surrounding Ndrova. We arrived at the island in the dead calm of a humid, overcast day. The water was the only refuge from the oppressive heat. Ronnie anchored the boat on top of the reef, and after

Trap containing two species of nautilus, Ndrova Island, Papua New Guinea, 1984.

hurrying into our gear, we plunged into the coolness of the tropical sea.

Reef walls of the tropical seas are one of Earth's great natural beauties. Sea fans extend outward from these vertical cliffs, waving gently in the slight currents. We dived about 80 feet into this still place, which was dark from the lack of sun but still crystal clear. We hovered against the wall, punctuating the darkness with staccato bursts of light from our great underwater strobes. I was so absorbed in photographing the vertical reef wall that I foolishly turned my back to the open sea behind me. Out of the corner of my eye I saw Ronnie stiffen and drop his camera. I had no doubts about what was happening. I felt a strong hand push me against the reef, and then heard a tremendous thud. Spinning around, I saw Ronnie holding his large underwater light, the front now broken, and a black-tip reef shark contorting wildly nearby. Two more sharks appeared from the deep blue in front of us. Our backs to the reef, we became creeping mollusks, moving upward and *out of there*. We scrambled

The rare species *Nautilus scrobiculatus,* Ndrova Island, 1984.

up onto the reef. Our boat had drifted off it with the tide, its anchor now dangling. In the quiet calm of the sea the boat floated a mere 30 feet from us, an easy swim on the surface. I looked at Ronnie and muttered a heartfelt thanks. He yelled to the Ndrova islanders, asking them to swim out for our boat. They laughed at us and paddled out in their dugout canoe to retrieve it.

Bruce Saunders remained in New Guinea for several more months and returned again several months later. During that time he trapped nautiluses around Manus, venturing as far as the island of Kavieng and also to Lae, a city on the northern coast of New Guinea. A rich story unfolded as the numerous data on shell diameters, color patterns, and variations among these separate populations began to emerge. The nautiluses from every distinct locality had their own diagnostic morphologies. Unlike the nautiluses from the more isolated islands such as Fiji, New Caledonia, and Palau, those from the New Guinea region showed greater variation. Clearly, interbreeding and admixture was taking place. The nautiluses of the outer Pacific islands, separated from the nearest populations by thousands of miles of deep ocean, showed little morphological variation. They had not exchanged genes with others of their kind in a long time. The nautiluses around New Guinea were a rich mélange of highly variable, interbreeding populations.

PORT MORESBY, 1984

After leaving Manus Island I traveled to Port Moresby, the capital of Papua New Guinea, to trap nautiluses in that region. Port Moresby should be one of the most beautiful cities on earth. It is situated next to a natural harbor, and its hills offer sweeping views of the sea that rival any found in Rio or San Francisco. But by the time I arrived, the beauty in Port Moresby had disappeared. It was a city under siege.

Port Moresby was being flooded by a tide of humanity. Refugees from West Irian and villagers from the forests arrived daily in search of a more promising future. The two main roads into the city run through giant garbage dumps on the outskirts of town. Shanty towns had sprung up in each of these dumps, supporting large populations of permanent residents who have found a livelihood in the city's garbage. In the city a long stretch of waterfront is lined with huts that sit over the water on stilts. On closer inspection this picturesque neighborhood quickly loses its charms. The toilets in these huts consist of a hole cut through the floor, and at low tide the waste of thousands of humans accumulates on the shore, creating an unimaginable stench.

By day the streets of Moresby bustled with the commerce and activity of a capital city. At night they were largely deserted, the turf of street gangs made up of youths from the city or new arrivals from the highlands. A decade after independence from Australia, Papua New Guinea and its capital city were still experiments in democracy. Moresby was one of the most expensive cities on earth; the price of a modest hotel room cost well over $100 per night. Most of the city's wealth belonged to the white ex-patriots who stayed on after independence. It was a city of fear, where the Papuan street gangs expressed their hostility toward the two-tiered economic system with a determined nocturnal campaign of violent crime against the whites. Martial law and a curfew were instituted in Port Moresby within a year of my 1984 visit.

I based my operations at the Motupori Research Station, a well-organized marine laboratory operated by the modern and uniformly excellent University of Papua New Guinea. The lab was run by a young British marine biologist, Dr. Nicholas Polunin, with the technical help of two American biologists, Dr. Pat Colin and his wife, Lori. My first few days were spent gathering the equipment and material I needed to construct nautilus traps. I also visited the headquarters of the Papua New Guinea Department of Fisheries and arranged with its chief to use a large 50-foot fishing boat and its crew. The only cost was the price of fuel and any overtime necessary for the crew, which was a real break. The lab had only small runabouts, and unlike the quiet calm of the Manus Island region, the sea off Port Moresby was constantly buffeted by 20-knot trade winds. And the water off the reef where we planned to trap was particularly rough. I was very happy to have a 50-foot boat under me in this turbulent sea.

We sortied out on our first morning of trapping in a brisk wind. The barrier reef lay several miles off the coastline, and by the time we arrived outside the reef, the boat was rolling unmercifully. I was very happy to get in the water. As we had done in other places in the Pacific, we tied our lines off on the reef wall, then carried the trap out to sea. When the 800-foot-long line was taut, we threw the trap off the end of the boat. We set three traps in this manner. My only problem was getting from the very rough sea back into the boat while dressed in full scuba gear.

The captain and crew of the fishing vessel were all Papuans. They were extremely efficient at their work and obviously pleased to be so successful with this new way of fishing. By noon we had set our three traps and headed back to Motupori Island, where we planned to spend the night before returning in the morning to collect them.

When we arrived back at the lab, the captain told me his crew needed food money for the following day. Also, since the next day was a Saturday, he explained, he and his crew were to be paid double, in advance, please. These guys had been very good, so I forked over the money. As my crew headed for the highway and Port Moresby, some 20 miles away, I yelled, "Tomorrow morning."

"Sure," replied the captain with a wave of his hand. "Seven

A.M." Iggy, the Papuan assistant at the marine lab, looked at the departing crew and laughed.

By nine the next morning I was really angry. The fishing boat rocked gently at anchor, and the crew was nowhere to be seen. About an hour later one crewman and the captain's mate stumbled into the lab compound reeking of whiskey, their eyes red, their heads obviously aching. "The Captain is sick today," one explained. I asked if he would come the next day. "No, I don't think so," was the reply. "The Captain is *very* sick."

I had use of the boat only for the weekend. The mate said he could run the boat, and after I pressed Pat and Lori into service, we set out after the traps.

The wind and waves were high and we wallowed heavily out to sea, finally arriving at the site of our tied-off traps. Because of the strong seas we had to keep the boat well off the dangerous reef. To raise each trap, I swam into the reef, untied the rope, and swam it back to the waiting boat. Once aboard, we pulled in the slack as the mate steamed away from the reef. The powerful winch rapidly pulled in the first trap. We watched with great excitement as it finally came into view. The trap contained nine nautiluses, all mature, and all small—only slightly larger than the specimens from Fiji. We had hit scientific pay dirt once again.

The second line was as successful as the first. This time 14 nautiluses were found. I had already exceeded my goal of 20. The happy mate smiled at our compliments and shared in our joy. He was performing admirably under rough conditions, and no doubt had visions of himself as a captain.

During the recovery of the third trap, things didn't go as well. I recovered the line and swam it back to the boat. Getting in and out was fatiguing in the heavy seas, and I was happy to be pulled on board with the end of the final line, which floated behind us. Then, for inexplicable reasons, the mate decided to imitate Captain Queeg: He turned the boat *toward* the reef rather than out to sea, and in so doing ran over the 3/4-inch polypropylene tow rope. The engine screamed in protest as the thick plastic line wrapped around the shaft. The mate quickly took the ship out of gear, but the damage was done. The prop was immobilized.

Now this was a problem. We were about 50 yards off the reef, but a strong wind was pushing us rapidly toward the surf breaking over its top. Anchoring was not an option because the reef wall was vertical. Once we were in water shallow enough to drop anchor, we would be on the reef. Our only hope was for me to dive down and try to remove the tangled rope.

Before going over the side, I quickly told the mate to keep his hands off the gear box while I was underneath the boat. The rope had made a fine macrame around the propeller shaft and rudder. I could see the reef looming ever nearer, and on board, panic must have been taking hold. With a sharp knife I was soon able to cut away most of the rope. Moments after I had finished, the mate decided to see if his prop was clear. To him it was all-important that he come home with the ship in one piece. I heard the gears clang into place; the turning shaft caused the few strands of rope still on the prop to pop loose. I was able to kick off to one side and avoid the spinning propeller, but the barnacle-covered bottom ripped the skin off the back of my legs. The rope was cut and all hope of getting the trap was gone.

I swam out to the boat trailing a cloud of blood, which made me prime shark bait. The mate was not in the least bit apologetic. I had gotten my nautiluses, hadn't I?

LIZARD ISLAND, AUSTRALIA, 1985

In June 1985 Bruce Saunders and I traveled to the Great Barrier Reef and set our traps in the passage Captain Cook had used to escape from the area. To our surprise we caught two kinds of nautilus: *N. pompilius,* which we had expected to see, and *N. stenomphalus*—the last missing species. The *N. pompilius* specimens were indistinguishable from those I had captured off Port Moresby. To me, the newly captured specimens of *N. stenomphalus* were the most beautiful nautiluses of all. Their shells were nearly pure white and their tentacles lined with lacy white frills, and the hood of each animal was covered with strange, tiny white denticles. We were ecstatic about our catch: Finally we had living proof of a nautilus species that formerly had been known only by its shell. We were later amazed to find that it was hybridizing with *N. pompilius*. One or the other of these species had migrated into this portion of the Great Barrier Reef, perhaps during the last ice age, and was now interbreeding with the species already there. In any case this presented major problems for us in understanding nautilus taxonomy. With two supposedly dis-

The first live *Nautilus stenomphalus* ever seen, Great Barrier Reef, Australia, 1985.

tinct species in the area, we were catching pure *N. pompilius,* pure *N. stenomphalus,* and hybrids with characteristics of each.

To make matters even more interesting, shells of the largest nautilus of all were washing up on beaches in the Great Barrier Reef region, suggesting that three species lived off that largest of coral reefs. Unfortunately we never succeeded in trapping one of these giants.

THE DIVERSITY OF NAUTILUS

In trying to figure out how many species of nautilus exist, the closest I've come to an answer is that there are probably far more than old Sowerby imagined. In the last two years alone two new, distinctive populations have been reported in addition to the previously unseen species Saunders and I trapped. In the Sulu Sea, west of the Philippines, a professional diving guide set traps and caught perhaps the most extraordinary nautilus yet known. About 30 tiny but fully mature live specimens were found on an isolated reef, far from any land area. Their shells were donated to science, giving Bruce Saunders and me the chance to study them. The adults were about four inches in diameter, even smaller than specimens from Fiji.

In 1985 an Australian fisherman who was shrimping on the continental shelf of Western Australia discovered the other extreme when he dredged up living nautiluses that were over a foot in diameter and larger than any previously known, including the biggest ones from Palau. These giants resemble the large drift shells found from time to time along the Great Barrier Reef.

The various nautilus species appear to fall into at least two biogeographic groupings. Nautiluses that live near islands separated from land by wide expanses of deep water constitute one such grouping, which includes the tiny specimens from Fiji assigned to *N. pompilius;* the giants of Palau named *N. belauensis; N. macromphalus* from New Caledonia; a form from the New Hebrides called *N. pompilius;* and most recently, a newly discovered form of *N. pompilius* from Samoa. The second biogeographic grouping of nautiluses consists of those that live on the block comprised of Australia and New Guinea.

We have learned much about the ecology and life habits of nautiluses. We know that they rarely swim more than a few feet above the sea bottom and that warm water temperatures—found at the sea's surface over much of the tropical Pacific—kill them. These two facts lead us to conclude that any long-distance migration by a nautilus must occur along the bottom of the sea. In addition to these limitations, nautiluses are further restricted by hydrostatic pressure, which causes their shells to implode at depths of around 2,000 feet, and by the nature of their buoyancy system, which appears incapable of keeping the chambers empty of liquid for sustained periods at depths greater than about 1,200 feet. According to a bathymetric profile map of the portion of the Pacific Ocean where nautilus lives, most island groups are separated by depths significantly greater than this. The nautilus's only hope for long-distance migration would be in the midwater region, at depths cooler than the lethal temperature limit but shallow enough to avoid implosion or inadvertent chamber flooding. Nautiluses, however, refuse to—or cannot—swim for long distances out of sight of the bottom.

Nautilus's reproductive strategy is poorly suited for colonizing new areas. Many invertebrates, including most sessile species, produce millions of gametes that are broadcast yearly into the sea. The newly fertilized larvae spend some time floating in the plankton at the sea's surface, and because of their microscopic size, they are easily transported by waves, wind, and currents. Many larvae can stave off metamorphosis—the time when they must leave the plankton, descend to the bottom, and begin adult life—and can potentially be transported across entire oceans. This type of larval dispersal gives the adults a wide geographic range. Corals use such a system, and many of the same coral species are present in various reef areas of the South Pacific. The reproductive system of nautilus is vastly different. Nautilus eggs are some of the largest known among invertebrates and may be as much as an inch and a half long. We have evidence that when a baby nautilus hatches from this large egg, its shell is already an inch in diameter. It spends no time in the plankton. Dispersal of the young, so common in many invertebrate species, appears to be nonexistent in nautilus.

The nautiluses on the isolated islands such as Palau, Fiji, New Caledonia, and Samoa appear to be completely cut off from others of their genus: They show almost no intraspecific variation. The males and females each have characteristic shell diameters that vary little from individual to individual, and gene flow with the other populations probably never takes place. How did these nautiluses get to their isolated habitats in the first place?

This question is a thorny one. Although none of the islands mentioned above is especially young geologically, we can safely assume that the nautiluses have not always been there. Some of these islands have excellent exposures of Tertiary-age sedimentary rocks, and there has never been a fossil found there that looked anything like *nautilus*. Somehow, and probably within the last five million years, nautiluses have come from elsewhere to colonize these islands.

The answer may lie partly in the size of the island groups themselves. Biogeographers have a name for rare, chance colonization of an isolated island: They say it has been colonized by a sweepstakes route. As improbable as any single chance migration may appear, the chances increase as the size of the island or island group increases. All of the islands listed above are large or form chains. Often nautilus cannot be found on smaller islands that are somewhat isolated from larger groups. It may be that these smaller islands are unsuitable for the nautilus's living needs. More probably, because of the small size of these islands, the chance migration event never occurred.

Time is an additional factor in sweepstakes-route colonization. When the time span is in the millions of years, even extraordinary circumstances become more likely. Each of these islands was probably colonized by the chance arrival of a pregnant female nautilus, or a pair of nautiluses brought by a typhoon or some other serendipitous event. Once a small population was created, natural selection would shape the shell's ultimate morphology, determining size, coloration, or coiling characteristics in response to the local environmental forces. When a few individuals start a new population, an evolutionary process called the "founder effect" becomes increasingly important. The small size of the Fijian nautiluses may be related to environmental conditions in which this feature is an advantage. On the other hand, the "founding" nautilus may have been an uncharacteristically small individual, carrying within its genes this heritable trait. A population starts from a very small number of individuals. The genetic information from each of them determines in large part the ultimate morphological characteristics of the population they eventually create.

The isolated islands show a suite of nautilus characterized by low variability and gene flow. In contrast, the Pacific continental coastlines hold a variety of morphotypes that seem to exchange gene flow. A *Nautilus pompilius* of intermediate size can be found from the southern regions of the Great Barrier Reef of Australia all the way around New Guinea, an aggregate coastline of many thousands of miles. With no deep barriers of water to prevent migration, genetic information may be slowly passing among the millions of nautiluses that surely live along this great coastline. Even the break between Australia and New Guinea, located at the Torres Strait, has a series of shallow reefs that can serve as stepping-stones to migration. The opposite of sweepstakes routes, such paths to migration are called corridors.

The Ice Age probably had a profound effect on the history of nautilus. When the glaciers were at their maximum advance, the sea level was far lower than it is today—perhaps as much as 450 feet lower. During these periods, which occurred four times in the last million years (and perhaps several times before that), migration to and from the various continents and islands within nautilus's known range was much easier than today. It is also possible that ocean water temperatures were lower, easing another obstacle to migration. Many of the places nautilus lives today may have been colonized in the last million years, during the Pleistocene epoch. I believe that this colonization began with nautilus populations that lived off Australia. About five million years ago nautiloids disappeared from the globe everywhere but around Australia. They then began the slow road back.

The diversity of nautiluses now known has caused us to revise what has been classically accepted about these creatures. Although the exact number of species is now a point to be argued by geneticists, there are more than were previously thought to exist. At two recent conferences about ancient mollusks, two respected scientists in two different disciplines reached the same conclusion: The genus *Nautilus* is in robust health, and certainly not declining toward extinction. Dr. David Woodruff, after inspecting the genetic material collected by Saunders and me, came to a conclusion identical to one reached by a paleontologist at the University of Kansas, Dr. Curt Teichert, whose inspection was based on the fossil record: Nautilus is a very young form, and in the midst of a great radiation of species. We may be entering a new age of chambered cephalopods.

Chapter 11
FERTILE EGGS AT LAST

The man who would finally complete Arthur Willey's quest lived his early life far from the sea. Bruce Carlson was born and raised in the Midwest, where winter winds blowing off the Great Lakes set him thinking about warmer places. A biology major, he finished college in the late 1960s and, recognizing the bleak prospects of employment in that field, joined the Peace Corps. His choice of country was never in doubt. Bruce Carlson packed his summer clothes and his new degree in a small suitcase and set out for Fiji.

Carlson finally realized his dreams in Fiji. He began to study the coral reefs surrounding his workplace, first learning the names and then the habits and biology of the resident fish and invertebrates that so fascinated him. He talked to the Fijians about the many undersea creatures and was intrigued by one animal in particular: nautilus. When he finished his tour in Fiji, Carlson moved to Hawaii and enrolled in a graduate program at the university. To finance his postgraduate education, he joined the staff of the Waikiki Aquarium.

Carlson was invaluable to the Waikiki Aquarium because of his wide knowledge, patience, and meticulous skill at keeping animals alive in the tanks. Though he rose rapidly through the ranks to become assistant director, he yearned to go west again, back to nautilus country. In 1976 he returned to Fiji, where he and I met for the first time, each of us trapping nautilus. Carlson managed to take a dozen living specimens back to the Waikiki Aquarium, where he kept them alive in cooled water for many months. Although nautiluses had thrived in the Nouméa Aquarium for years, the first exhibition of live specimens in Hawaii created a public sensation. When

Carlson received money to expand Waikiki's nautilus collection, he decided to go for the largest ones then known: the nautiluses in Palau.

Carlson started to make regular trips to Palau, and acquiring nautiluses for public display soon led him to develop an interest in the animal's scientific aspects. Carlson masterminded the first successful tracking of nautiluses in Palau, and was the first to realize that temperature change played a major role in the creatures' daily migrations.

In 1976 Carlson discovered that the nautiluses stocked at the Waikiki Aquarium were laying eggs. Ecstatic, he thought that he would soon have baby nautiluses. But the incubating eggs sat quietly, until they finally rotted away. Like Arthur Willey, René Catala, and Arthur Martin before him, Carlson collected, cared for, and ultimately discarded infertile nautilus eggs.

Why were the nautilus eggs in aquariums always infertile? Perhaps the great difference between the pressure in the aquarium and the deep-water reef habitat was the answer. But nothing could be done about low pressure in the aquarium. Constructing tanks pressurized to 40 atmospheres would cost a fortune. Willey had also worried about pressure and had tried to overcome the problem by putting the nautiluses and their eggs in cages and sinking them to their natural depth, but even this failed.

Bruce Carlson had breeding specimens, outstanding facilities, and plenty of time. He was determined to be the first to propagate nautilus in captivity. He couldn't do anything about pressure, so he decided to manipulate the only other variable he could control: temperature. Carlson changed the temperature in the nautilus aquarium twice a day, simulating the fluctuations the animals experienced in Palau during their twice-daily depth changes. Each night he raised the water temperature and each morning cooled it. The nautiluses began to lay eggs, but again the eggs were seemingly infertile.

By March 1985 Carlson had examined several hundred nautilus eggs. He passed on yet another batch of 20 to his colleague John Arnold of the University of Hawaii. The eggs sat undisturbed in an incubator for five months until Arnold opened them and found, to his astonishment, a tiny, caplike shell. Out of 20 eggs, four living embryos in various stages of development were recovered.

More and more living embryos were found in the eggs coming from the nautiluses undergoing daily temperature change. The developing embryos were extraordinary; they sat atop a huge mass of yolk, slowly building seven tiny chambers into their shells. The entire egg interior, including yolk and embryo, rotated every two minutes. The embryos grew very slowly, taking perhaps a year to develop inside the eggs. Arthur Willey's quest was finally completed.

Postscript

THE PHILIPPINE ISLANDS, 1987

My first real view of the Tanon Strait came from the window of an Air Philippines plane as we made our final approach into Dumaguete City. Long and narrow, bounded by the islands of Negros to the west and Cebu to the east, this great body of water has always been the world's chief source of nautilus shells. For hundreds, and probably thousands, of years Filipinos have fished this warm inland sea with their intricately made wicker traps. Usually the catch was deep-water fish and crabs, but often their traps held a strange, swimming shell with many tentacles that they called *lagang*. The Filipinos used to throw these wondrous creatures back into the sea because they considered the flesh inedible. Only in the twentieth century, when curious foreigners started to offer good money for these large brown-and-white shells, did fishermen begin to keep them.

I went to the Tanon Strait to see firsthand the most well known and heavily fished population of nautilus on earth. Anyone who has held a nautilus shell has probably touched a bit of the Tanon Strait. Through geological accident this body of water has become the ideal habitat for nautilus. Elsewhere in their range, nautiluses live on the seaward sides of steeply dipping reef walls, clinging to the narrow habitable band between lethally warm surface water above and implosion depth of 750 meters. The Tanon is 500 meters at its deepest, and most of its gently sloping bottom is in a depth zone of 100 to 300 meters, which is ideal for nautilus. These large nautiluses, which we assign to the species *N. pompilius,* can freely swim from side to side of the 30-kilometer-wide strait and roam its entire 230-km length. But the nautiluses of the Tanon are probably cut off from other nautilus populations in Philippine waters. Both ends of the strait are closed by shallow sills at depths of less than 100 meters. Because the nautiluses cannot tolerate the very warm water temperature found in the regions of these sills, they are effectively landlocked.

The mild currents and calm weather of the Tanon Strait add to its reputation as an ideal habitat for nautilus. The seas surrounding most of the Philippine Islands experience large daily tidal changes that, in turn, generate powerful tidal currents along most of the coastline. To the north and south of the Tanon Strait, the raging tidal currents of the Visayan Sea create hazards for marine life and for the Filipino fishermen in their small outrigger canoes. Within the Tanon Strait, however, currents are greatly reduced because of the barred entrances at either end. Winds are equally light. Located at latitude 10 N, the Tanon straddles the equatorial doldrums and is thus not plagued by cyclones or trade winds. During much of the year the Tanon Strait is a huge, flat, sweltering lake. Beneath its placid surface, the organic-rich muds of the bottom became an ideal habitat for a diverse lot of crustaceans and fish. Over time, the cool waters beneath the strait's 100 meter thermocline filled with slowly multiplying nautiluses.

Many generations of perhaps the world's cleverest fishermen have lived along the palm-fringed shores of the Tanon Strait. Blessed with a bountiful sea and fertile volcanic soil, this area is truly one of the Earth's paradises. But as people multiplied along these sun-drenched shores, the seemingly inexhaustible supply of fish began to dwindle, and with the disappearance of favored species, the people of the Tanon were forced to eat once-scorned varieties. Granted, there were certain advantages to the changing biota of the Tanon Strait. Sharks, for instance, became increasingly rare, as did seagoing crocodiles, many of which were turned into food and shoes. But the large predators were not the only species to vanish. During the 1960s and 1970s the numbers of traditionally caught fish and other marine life dwindled. Gone were the large, proud tuna and the sea turtles. Giant clams, once so numerous along the continuous reefs lining the strait, had also been fished to extinction. The land suffered much the same fate. The rich rain forests were cut back to a fraction of their original area. Rice fields replaced lowland forests, and the pesticides used on these fields found their way into the streams and ultimately the sea. The birds were soon victims of this chemical onslaught. I had never seen an ocean without seabirds until I came to the Tanon Strait.

To maintain a steady supply of fish for the burgeoning population, new means of catching them were introduced. The Nobel Company built dynamite plants in the area, and dynamite fishing was born. Encouraged by the government of Ferdinand Marcos, fishermen found that if they dropped several sticks of dynamite onto a reef, they reaped a harvest beyond their wildest dreams. No one seemed too bothered that the reef was blown up as well; after all, there were thousands of miles of coastline and endless reefs. But in 20 years virtually all of the coral reefs lining the Tanon Strait had been devastated. And as the reefs disappeared, so too did the breeding grounds for many species of fish. The people of the Tanon had to resort to eating ever smaller species and catching the juveniles of the larger ones. But catching lots of little fish was hard work, and in the 1970s came another novel method—cyanide fishing. The fishermen found that concentrated cyanide dumped into the sea brought up lots of fish. Some were edible.

Long before the fall of President Marcos in 1986, fishermen began to realize that these "modern" methods may not have been such good ideas, because as breeding sanctuaries and reefs disappeared, fish catches diminished sharply. In the mid-1980s the government declared these practices illegal, but it had no way of enforcing the new conservation laws. Faced with starvation, the fishermen continually returned to their dynamite and cyanide. Conservationists made an effort to halt fishing in some areas and tried as well to stop the clear-cutting of rain forests. But often these well-intentioned men became targets of the political far right, which used violence against those attempting to stall "development" of the Philippines.

I saw the wreckage of the reefs during scuba dives along the shattered coastline of Negros Island. By the time of my visit, everyone including the government recognized that this wholesale destruction had changed the nature of the marine biota. On the day I arrived, a small article in the *Manila Bulletin* reported: "Destruction of coral reefs due to dynamite blasting, pollution, and other fishing methods has drastically reduced annual fish production. A study by the Bureau of Fisheries indicated that more than 50 percent of the marine reef centers are in poor condition, and are on the verge of extinction. Fish catch has decreased 48 percent over the last ten years." Unfortunately, while the fish catch was plummeting, the Philippine population was skyrocketing, and continues to do so. Today nearly 60 million people live in the Philippine Islands. At the current birth rate, the population will double every 28 years.

I was joined in my journey to the center of the Tanon Strait's nautilus fishery by Angel Alcala, the director of the marine laboratory of Dumaguete's Silliman University. Dr. Alcala is a Stanford-educated herpetologist who has preached conservation and safe fishery practices for 20 years. He was then involved in direct action. His laboratory, home for many visiting scientists such as myself, was engaged in numerous conservation and fishery replenishment projects. I was amazed by the crocodile-rearing pens, the tanks of giant clams, and the shrimp aquaculture studies under way there. Alcala had pioneered the practice of building artificial reefs to replace the destroyed natural reefs of Negros, and already his pilot projects were showing results in slowly increasing fish yields. He has replanted giant clams along the shores and—most heretical of all—has designated several reserves along the coast and on small offshore islands in an attempt to save a few scraps of the natural coral reef. Such activities have not earned him any friends among the political right wing, which seems to covet a monospecific ecosystem for the Philippines—*Homo sapiens* only.

After two hours of driving on tortuous roads, we reached the tiny village of Bindoy. We stayed in the home of Wilson Vailoces, and for a week we dined on the most exquisitely prepared rice and fish, served in every way imaginable. Wilson was a local minister as well as the most respected fisherman in the area. For years he had also served as a guide and specimen procurer to the scientists visiting Bindoy to study nautilus.

Wilson had his first association with a scientist in 1971. For one year he set traps each night, retrieved them the next morning, and took the captured nautilus to Noreen Haven, a brave and determined researcher from British Columbia's Simon Fraser University. Not even the redoubtable Arthur Willey had discovered a spot he could stay in for so long that was so rich with nautilus. Haven's goals were much the same as Willey's; she hoped to acquire developing nautilus embryos and to better understand the mysterious reproductive habits of this enigmatic cephalopod. Every morning Wilson would return from his traps with about a dozen of the Tanon's large nautiluses. Haven then dissected them and examined the size of their reproductive organs in order to determine whether nautilus bred year-round or during only one season. Like Willey, she put mated pairs of nautiluses back into the sea in closed-off traps, hoping that they would lay eggs. Like Willey, she never succeeded in this endeavor. Every day for a year, Wilson would row out to these traps and feed the uncooperative specimens. As the year went by, the hundreds of data points, each representing another nautilus, turned into thousands. By the end of the year, well over 3,000 nautiluses had been caught and studied. When I met Wilson, he showed me a yellowing ledger book with meticulous records documenting this yearlong trapping effort. Each trap left in the sea became a carefully noted experiment.

The people of Bindoy still have kind memories of Noreen Haven. She lived simply, in the Filipino way; she ate their food and followed their style. Her researches resulted in three papers, one documenting three months' worth of her catches, with observations on behavior and feeding, and the other two documenting her studies on the anatomy of the reproductive system. Although these papers advanced our understanding of nautilus, they exacted a terrible price on the Tanon population. In 1971 the local fishermen of Bindoy had watched in amazement as one small man in a tiny outrigger canoe, using only three traps, brought in 3,000 nautiluses that year. Some of these fishermen began to build traps, and soon there were many more than Wilson's three dropped into the placid waters off Bindoy each night.

Word of the rich nautilus catches came to the eyes and ears of other scientists, and in 1975 Noreen Haven returned to the Tanon Strait with a group of specialists eager to study nautilus at Bindoy. They came abroad the research vessel *Alpha Helix,* which was operated by the U.S. National Science Foundation. The local fishermen soon heard that the visiting scientists were paying the exorbitant price of more than a dollar a nautilus. It wasn't long before the scientists were swamped with specimens. Unfortunately the ship was not equipped with a means for cooling seawater in which to keep the captured animals alive. Although intricate experiments on respiration and metabolism were attempted, the results had to be viewed skeptically because of the poor health of the specimens, which suffered in the lethally warm seawater aboard the ship. According to Haven, 220 nautiluses were sacrificed during this expedition.

The *Alpha Helix* returned in 1979 with a new batch of scientists, and once again they were turning away freshly captured specimens offered by the local Bindoy and Cebu fishermen. In 1980 and 1981 Bindoy was visited by a group of scientists from Japan. The Japanese tried their own trapping at first, and caught only one specimen. So the call went out again to Wilson and his fellow fishermen and the nautiluses came in. But these Japanese scientists were not alone in their curiosity, nor were they the major contributors to the killing of nautiluses: Every time someone in America, Japan, or Europe bought a nautilus shell, the Bindoy fishermen went out to their traps for more. Because nautilus fishery is completely unregulated, no records were kept, but Wilson estimates that a minimum of 5,000 animals were caught each year in Bindoy solely for the shell dealers. Until the last few years, that is.

I arrived at this center of nautilus activity in 1987. The local fishermen heard of my desire for fresh, living specimens for study, but they were a bit disappointed that the maximum I wanted was 20. But money is money, and the fishermen went to it. First they had to build new traps and repair the old ones, which lay scattered about, cobweb-encrusted and rotting. This should have alerted me.

We spent our first afternoon in the shade of Wilson's house, moving slowly in the thick heat as we prepared the traps for the evening's set. Fresh chickens were slaughtered for bait and meticulously cubed and strung into the traps. The woven traps were weighted with rocks and the lines spliced and attached. By 5:00 P.M. we were ready. We were only one of ten trap-laden boats that paddled out onto the unruffled surface of the Tanon Strait. Wilson took us to his favorite nautilus fishing grounds, located above a 200-m-deep plateau. He said it was the richest site in the Tanon for catching the animals. The various boats to the north and south of us reached their positions, and forty wicker traps sank into the sea. Never before had I approached the sea with so many traps. My misgivings all had to do with the number of excess nautiluses I would have to pay for and then throw back the next morning. The fishermen wanted to be paid for each nautilus they caught whether or not I kept it, and they thought I was crazy to even think of returning perfectly good specimens to the sea.

We returned to Wilson's house and ate our meal of fish (very small fish, called teeny-weeny crunchy see-throughs by the locals). I spent a typical nautilus fisherman's night of anxiety about the traps, burrowing under the damp sheet of my bed in an attempt to avoid the swarming mosquitoes. At the first hint of dawn, Wilson rowed us out unerringly in the dark to the bamboo buoy. We took turns hauling in the 600 feet of rope with the heavy traps on the end. As the first one came to the surface, in a dead heat with the rising sun, I saw my first living nautilus from the Tanon Strait. But our catch was a disappointment, for in three traps that was our only specimen. Wilson was deeply ashamed. He was the acknowledged master. When he had fished for Noreen Haven, this spot had averaged about five nautiluses per trap each night. I wasn't unduly concerned, since I knew that at that moment 37 other traps were coming to the surface, undoubtedly carrying the precious creatures. Amazingly, when all of the fishermen were assembled, we found that our single nautilus was that night's only catch.

For a week we fished the 40 traps in the Tanon, moving them shallower and deeper and north and south of Bindoy. In 1971 some 40 traps fished for a week might have yielded approximately 1,400 nautiluses. We caught just three specimens in one week.

Dr. Alcala and I moved to Cebu and found the same story. Gradually the sad truth became apparent. We found rotting and disused traps everywhere. Over the last several years the fishermen had gradually stopped trapping for nautilus in the Tanon because of poor or nonexistent catches. Nautilus had disappeared from the Tanon Strait. I thought back in horror to the three I had caught. Could they have been the last living nautiluses of this once-rich population?

Viewed in hindsight, it is not surprising that this crash took place in the Tanon Strait's nautilus fishery. We know that nautiluses take as long as 15 to 20 years after hatching to reach reproductive maturity, and then produce very few young (perhaps as few as 10 each year). What is surprising is that Noreen Haven found that only about 10 percent of the population she sampled were females. The situation made me think of a story by the brilliant English writer John Wyndham in which creatures inhabiting the deep sea climb into pressurized machines to "fish" the land for people. Our reproductive abilities and life span are remarkably similar to those of the nautilus. How long would it take a determined group of "fishermen" to wipe out a human population of 100,000 at a rate of 10,000 per year, even with our best reproductive efforts to make up the losses?

Nautilus is, perhaps, an animal easily driven to extinction. After all, the geological record is littered with cephalopod extinctions. Again and again lineages of nautiloids or their look-alike cousins, the ammonoids, vanished from the Earth in a short time. But the extinction of the Tanon nautilus population (if extinction it is; I hope a few remaining animals will repopulate this area) was unique because it was caused by man. Perhaps it was the fishing, or the destruction of the reefs where the nautiluses probably bred, or the loss of fish, or pollution from the poisonous copper mines that line the northern Tanon Strait. Whatever the reason, a tragedy has occurred. More than a million people now live along the shores of this region. Perhaps that many nautiluses once lived there.

When I was six years old my mother bought me a nautilus shell that probably came from Tanon Strait. She didn't want to, but I kept whining until she did. It cost only a couple of dollars. Years later, at the end of scholarly or popular lectures about nautilus, I would often face the familiar question, "Aren't nautiluses rare?" I would always answer, "Anything that costs only a couple of dollars is not rare or on the verge of extinction."

Acknowledgments

This book was written with the help of many friends and colleagues. I would especially like to thank those who patiently answered my questions, including Anna Bidder, Arthur Martin, Eric Denton, John Gilpin-Brown, Irene Catala-Stucki, and Claude Spinosa. Much of the background research, conducted in England at Cambridge University and Plymouth Laboratory, was aided in no small way by Dr. Martin Wells and Amanda Reid. I was supported greatly by the office and technical staff of the Department of Geological Sciences at the University of Washington, who did a magnificent job of finishing the typing and the photographs with skill and speed.

INDEX

(Page numbers in *italics* refer to captions.)

Picture Credits

Page 13: From *The Extinction of the Ammonites,* P. Ward. © 1983 by Scientific American, Inc. All rights reserved.

Pages 14, 15: From *The Buoyancy of the Chambered Nautilus,* Ward, Greenwald and Greenwald. © 1980 by Scientific American, Inc. All rights reserved.

Pages 35, 50, 51, 59: From Willey's Monograph, 1902.

Pages 48, 68, 73, 84, 114, 116, 117, 151, 154, 156, 166, 205, 206: P. Ward.

Pages 63, 75: Courtesy of A. Bidder.

Pages 82 (top): John Gilpin-Brown.

Pages 82 (bottom): Courtesy of Eric Denton.

Page 85: *The Journal of Marine Biological Association,* Eric Denton and John Gilpin-Brown, 1961. Reprinted with permission.

Page 109: *Le Memorial Caledonien,* Noumean Diffusion, New Caledonia. Reprinted with permission.

Pages 131, 132: *Paleobiology,* photos by P. Ward, 1981. Reprinted with permission.

Page 136: *Nature,* photo by P. Ward, 1983. Reprinted with permission.

Page 143: *Science,* W. Saunders and C. Spinoza. Reprinted with permission.

Page 167: From *The Journal of Marine Biological Association*, photo by P. Ward. Reprinted with permission.

Pages 168, 169, 170: *Veliger,* P. Ward, 1986. Reprinted with permission.

Page 177: *Veliger,* photo by P. Ward, 1981. Reprinted with permission.

Pages 182, 187: B. Brumbaugh.

Page 189: *Nature,* P. Ward, 1984. Reprinted with permission.

Pages 192, 199, 200: Courtesy of P. Ward.

Page 211: Tom Landry.